Linear CMOS RF Power Amplifiers

Hector Solar Ruiz · Roc Berenguer Pérez

Linear CMOS RF Power Amplifiers

A Complete Design Workflow

Springer

Hector Solar Ruiz
Roc Berenguer Pérez
Electronics and Communication Department
Centre of Technical Research (CEIT)
and University of Navarra (Tecnun)
San Sebastian
Spain

ISBN 978-1-4614-8656-5 ISBN 978-1-4614-8657-2 (eBook)
DOI 10.1007/978-1-4614-8657-2
Springer New York Heidelberg Dordrecht London

Library of Congress Control Number: 2013945144

Printed on acid-free paper

Springer is part of Springer Science+Business Media (www.springer.com)

You have had the good fortune to find real teachers, authentic friends, who have taught you everything you wanted to know without holding back. You have had no need to employ any tricks to steal their knowledge, because they led you along the easiest path, even though it had cost them a lot of hard work and suffering to discover it… Now, it is your turn to do the same, with one person, and another—with everyone

Saint Josemaría Escrivá
Founder of the University of Navarra
To our families

Preface

The great spread of wireless technologies that is now observed reflects the interest in greater connectivity and pushes the development of portable devices that are able to connect to these emerging wireless technologies. Portable devices, then, need to offer increasing connectivity capabilities while maintaining their performance in terms of size and autonomy. Therefore, portable devices must further reduce not only their power consumption but also the size of their electronics. In other words, high performance, low cost, and highly integrated Radio Frequency Integrated Circuits (RF ICs) are increasingly required by the consumer electronics industry.

CMOS integrated technology has played an important role in this wireless explosion due to its high functionality, integration capabilities, and low cost. Consequently, power amplifiers (PAs) implemented in standard CMOS processes, which offer performance close to that found in more expensive technologies, such as GaAs, are highly attractive. This is not only because CMOS technologies are extensively currently used in RF ICs implementations but also because CMOS PAs may offer low cost and high integration characteristics.

However, the PA is still an RF component that has not been completely integrated within the whole transceiver due to the existing trade-off between high-performance and high integration characteristics. If high-performance PAs are required, designers focus on expensive processes that prevent PAs from being implemented in low cost, highly integrated devices. Conversely, if high integration is desired, achieving high-linearity and high-efficiency CMOS PAs is still a challenge. In addition to this, PAs have a direct impact on transceiver performance because the PA power consumption may easily make up 50 % of the overall power consumption of the transceiver, meaning that a high-performance PA is crucial.

This work, then, focuses on design techniques for high-performance, fully integrated linear CMOS PAs for wireless applications. The work provides a complete flow for the design of the CMOS PA by describing the steps from the very beginning of the design process. The book provides an overview of the metrics that quantify the performance of the PA in order to obtain the PA requirements. In Chap. 3, the linearity and efficiency metrics of PAs can be found along with the metrics of PAs that handle digitally modulated channels. Stability and power capability parameters have been also included.

Once the specific requirements of the PA have been established, this work provides designers with a PA model to help anticipate the expected performance of the PA. Based on the most important design parameters such as biasing, supply voltage, inductor quality factors, current consumption, etc., the model provides the expected metrics of the PA in terms of efficiency, linearity, and output power levels. This model proves, then, to be a useful starting point in the first design steps. The model description can be found in Chap. 6.

Once parameters such as current consumption, supply voltage, or the required inductor quality factors have been quantified, the book discusses the optimization process of all the PA stages. Based on a linear CMOS PA example, the output matching network, output, and driver amplifying stages, the input matching network, and the interstage matching network are detailed. In addition, the main issues that must be considered in the PA layout design process in order to avoid performance degradation are also presented. Special attention is paid to the issues of integrated inductors in PAs, along with the extra considerations that designers must know in order to optimize inductor performance. The details for the PA and the integrated inductor optimization process are also found in Chap. 6.

The test setups and procedures required to characterize a PA are described in Chap. 7. In order to fully characterize PA performance, single-tone test and tests based on digital channels should be performed. The book presents both types of tests, along with results based on the aforementioned linear CMOS PA example. In addition, test setups and procedures for measuring inductors for PAs are included.

The book also provides an introductory overview of the impact of the PA in the transceiver quantified for modern communication standards in Chap. 1. Chapter 2 addresses the issues and limitations that CMOS processes impose on the design of high-performance linear PAs such as the low supply voltage that is available in modern submicron CMOS processes or low transistor transconductance. This chapter also details several other aspects of CMOS processes, such as substrate losses, impedance transformation, or stability and reliability issues.

Fundamentals of PAs, i.e., the classification of PAs into different classes, as current source or as switch-type PAs, are also presented in Chap. 4. The practical uses of the different PA classes in implementing a linear PA architecture are also presented; examples might be using class C PAs in linear Doherty PAs or combining class D PAs to implement an outphasing PA architecture.

Finally, Chap. 5 is devoted to PA architectures that are of interest for building fully integrated PAs in order to achieve higher output power levels, enhanced linearity, or better efficiency. Power combined PAs and the Doherty architecture, along with dynamic supply, adaptive biasing or digital predistortion techniques, and the use of cascode transistors are all interesting solutions to boost PA linearity, efficiency, or output power levels in CMOS processes with limited supply voltage.

Contents

Acronyms

3GPP	The 3rd Generation partnership project
ACLR	Adjacent channel leakage ratio
ACP	Air coplanar probe
ADS	Advanced design system
AM	Amplitude modulation
BPSK	Binary phase shift keying
CG	Common-gate
CMOS	Complementary metal oxide semiconductor
CS	Common-source
DAT	Distributed active transformer
EIRP	Effective isotropic radiated power
EER	Envelope elimination and restoration
ET	Envelope tracking
EVM	Error vector magnitude
FCC	Federal communications commission
GMSK	Gaussian minimum shift keying
GSG	Ground-signal-ground probe
GSM	Global system for mobile communications
HSUPA	High-speed uplink packet access
IC	Integrated circuit
IF	Intermediate frequency
IP3	Third order intercept point
ISS	Impedance standard substrate
K	Rollett stability factor
LFA	Low-frequency amplifier
LTE	Long-term evolution
LUT	Lookup table
MIM	Metal-insulator-metal
NMOS	N-channel metal-oxide semiconductor
OFDM	Orthogonal frequency-division multiplexing
OFDMA	Orthogonal frequency division multiple access
PA	Power amplifier
PAE	Power-added efficiency
PAPR	Peak to average power ratio

PCT	Parallel combining transformer
PDM	Pulse density modulation
PM	Phase modulation
PMOS	P-channel metal-oxide semiconductor
PSCT	Parallel-series combining transformer
PWM	Pulse width modulation
PWPM	Pulse width, pulse position modulation
Q	Inductor quality factor
QAM	Quadrature amplitude modulation
QPSK	Quadrature phase shift keying
RBW	Resolution bandwidth
RF	Radio frequency
SAW	Surface acoustic wave
SC-FDMA	Single carrier frequency division multiple access
SCT	Series combining transformer
SGN	Signal generator
SGS	Signal-ground-signal probe
SOI	Silicon-on-insulator
SOLT	Short-open-load-thru
SOS	Silicon-on-sapphire
SPA	Spectrum analyzer
TDDB	Time dependent dielectric breakdown
VBW	Video bandwidth
VNA	Vector network analyzer
VSA	Vector signal analysis
WCDMA	Wideband code division multiple access
WLAN	Wireless local area network
WMAN	Wireless metropolitan area network
WPAN	Wireless personal area network
ZVS	Zero voltage switching

Chapter 1
Introduction

Abstract This chapter introduces the power amplifier (PA) as a component within the transceiver. It then moves to a discussion of the impact of the PA in terms of power consumption and the new requirements of modern wireless communication standards, and it ends with a description of the importance of the peak to average power ratio (PAPR) for PA performance and the issues related to CMOS (Complementary Metal Oxide Semiconductor) processes for PA implementation.

The Power Amplifier

The power amplifier is the last component of the transmission chain, placed just before the antenna. Figure 1.1 illustrates a superheterodyne transmitter. It can be seen that the signal at the base band is first upconverted to an Intermediate Frequency (IF), filtered through a selective filter (SAW), and then IF amplified. After that, the signal is again upconverted to the final RF frequency and filtered. Normally, there is a PA driver that performs an initial amplification before it arrives at the PA. Therefore, the PA performs the final amplification of the transmitted signal so the signal can be received at the required distance and with the desired quality. As the PA is the last component in the transmission chain, it must deal with the highest power levels. Consequently, the PA usually shows the greatest power consumption in the transmission chain, which means that the efficiency of this chain could be practically reduced to that of the PA. Furthermore, this component strongly influences output signal quality, which is greatly affected when the PA works close to its nonlinear performance.

Impact of PA on Integrated Transceivers

The ratio of the PA's power consumption within the wireless transceiver has always been considered high. However, modern wireless standards can be very different: channel bandwidth or channel modulation, frequency band or output

H. Solar Ruiz and R. Berenguer Pérez, *Linear CMOS RF Power Amplifiers*,
DOI: 10.1007/978-1-4614-8657-2_1,

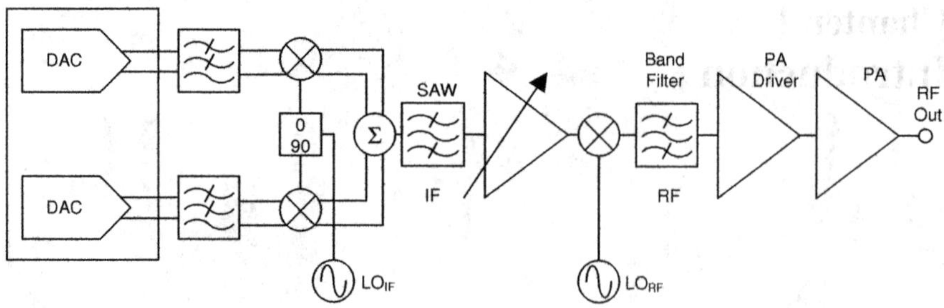

Fig. 1.1 Transmitter components diagram

power levels differ from one standard to another. Therefore, the impact of the PA on the transceiver will vary, depending on the target application. For that reason, it is very useful to quantify and detect which aspects of the final standard the designer must focus on when tackling the design of the PA.

Requirements of Modern Wireless Standards

There is a plethora of different parameters that describe modern wireless standards and the number keeps increasing with the appearance of new standards. However, there are two main system parameters that affect the performance of a PA: communication range and channel spectral efficiency. These two parameters directly fix the other two main aspects of a PA: the linear output power and PA efficiency.

Effect of the Communication Range

Figure 1.2 shows a simplified block diagram of a generic RF transceiver. There are five major circuit building blocks on the left side of the diagram: the transmitter front-end is responsible for modulation and up-conversion; the receiver front-end is for down-conversion and demodulation, the transmitter and receiver base band blocks are for signal processing, and the synthesizer generates the required carrier frequency. To the right of these blocks, the power amplifier block amplifies the signal to produce the required RF transmit power to the antenna and can be either integrated into the transceiver or be external. The power consumption of the transceiver will therefore comprise the power consumption of all these blocks.

From this simplified scheme it is possible to quantify the impact of the PA in the transceiver for actual wireless communication standards. In order to make that calculation, a simple definition of the impact of the PA is offered in (1.1).

$$PA\_impact = \frac{P_{PA}}{P_T} \tag{1.1}$$

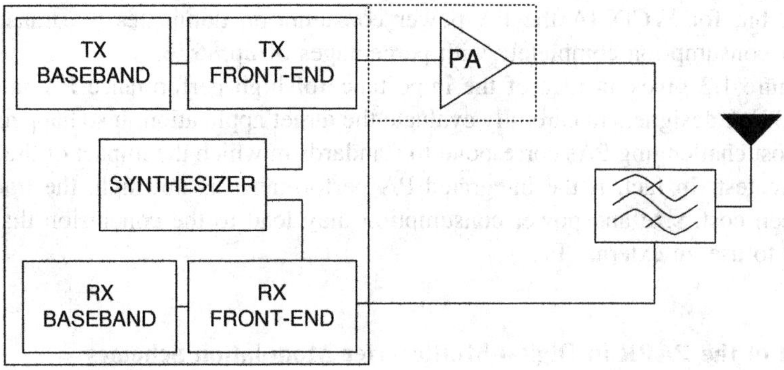

Fig. 1.2 Transmitter simplified block diagram

Where P_{PA} is the power consumption of the PA and P_T is the power consumption of the whole transceiver, including the PA.

It is now possible to apply this definition to several implemented transceiver designs for three different standards in order to quantify the PA impact. Bluetooth, 802.11g and 2 GHz WCDMA transceivers have been chosen as they cover the different communication ranges: the WPAN networks of Bluetooth, the WLAN networks of 802.11g and the WWAN networks of WCDMA.

The characteristics of the transceivers are in Table 1.1. All the transceivers are implemented in a CMOS process and within each standard the transceivers show similar output power values.

It must be noted that in the case of 802.11g and especially WCDMA, only a few transceivers integrate the PA. For that reason the same external PA has been applied to transceivers of the same standard. These external PAs are also implemented in CMOS and yield very good and realistic results; they are shown in Table 1.2.

As Fig. 1.3 shows, the impact, although significant, is not the same for each standard and clearly depends on the output power levels. For Bluetooth it is around

Table 1.1 Performance of state-of-the-art of CMOS transceivers for bluetooth, 802.11g and WCDMA

Standard	Trans.	P_T (mW)	P_{PA} (mW)	Impact (%)	Trans. P_{OUT} (dBm)	CMOS
Bluetooth	[1]	19.5	7.5	38.5	0	0.25 um
	[2]	123	55	44.7	2	0.18 um
	[3]	92.13	33.33	36.2	2	0.13 um
802.11g	[4]	1144	690	60.3	−4	0.18 um
	[5]	1212	690	56.9	−3	0.18 um
	[6]	1249	690	55.2	−4	0.18 um
WCDMA	[7]	2266.4	1700	75.0	4	0.13 um
	[8]	2114	1700	80.4	6	0.13 um
	[9]	1969	1700	86.3	0	90 nm

40 %, but for WCDMA the PA power consumption dominates the transceiver power consumption completely with percentages of up 86 %.

Figure 1.3 gives an idea of the importance of high performance PAs and the need of PA designers to carefully evaluate the target application. It so happens that the most challenging PAs correspond to standards in which the impact of the PA is the greatest. In fact, if the integrated PA performance is not high, the trade-off between cost, size and power consumption may lead to the conclusion that it is better to use an external PA.

Effect of the PAPR in Digital Multicarrier Modulation Schemes

As mentioned previously, PA performance is mainly affected by the continuous need for higher data rates in limited channel bandwidths, i.e. higher spectral efficiencies. A paradigmatic example can be observed in the evolution of 3GPP standards for mobile communications, shown in Fig. 1.4. It is clear that there is an increasing requirement for a higher bit rate, from the 9.6 Kbps channel data rates of the 2G standards to the 50 Mbps for the uplink in LTE in Release 8. This continuous need for higher bit rates also has an impact not only on the channel bandwidth, which ranges from 200 kHz for 2G to 20 MHz for LTE, but also on modulation, where 2G uses constant envelope GMSK modulation whereas LTE

Table 1.2 Performance of state-of-the-art CMOS PAs for 802.11g and WCMA

Standard	PA	P_{OUT} (dBm)	PAE (%)	CMOS
802.11g	[10]	21.2@EVM = −28 dB	19	65 nm
WCDMA	[11]	28@ACLR = −35dBC	36.5	0.18 um

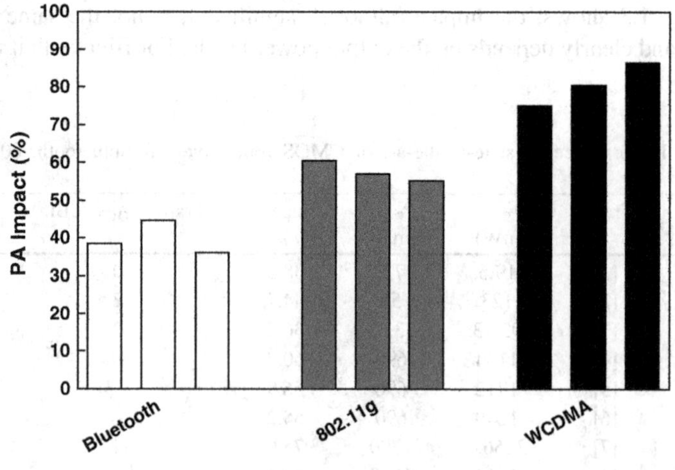

Fig. 1.3 Impact of the PA power consumption on three different wireless communication standards: Bluetooth, 802.11g and WCDMA

permits multicarrier channels with 16 QAM or even 64 QAM modulation for the uplink data transmission in order to increase spectral efficiency. High spectrally efficient channels require, then, more complex modulation schemes accompanied by high PAPR modulated channels.

In fact, as Fig. 1.4 illustrates, the PAPR has undergone a clear increase in the different 3GPP standard generations: from 0 dB for GSM to 8.5 dB for LTE [12–14].

Signals with a high PAPR indicate that at certain moments the transmitted signal may have peak power values that are much higher than the average. This leads to the necessity of using amplifiers with highly linear characteristics, otherwise the excessive clipping of the signal would lead to a distortion of the transmitted signal and out-of-band radiation; needless to say, these two phenomena are limited by standards in terms of maximum limits for the error vector magnitude (EVM) and adjacent channel leakage ratio (ACLR).

However, the trend of higher PAPRs seems unavoidable in wireless standards and in fact is a big concern for designers and standardization bodies. For example, although it exhibits a high PAPR, LTE uses SC-FDMA access for the uplink in order to mitigate the even higher PAPR from the OFDMA technique used in the downlink channels.

The effect of high PAPRs on the PA is direct in terms of the linearity requirements of the PA and also in terms of the PA efficiency. Maximum efficiency is normally achieved by the PA at maximum power, but a high PAPR transceiver will reach this power level for only a small amount of time. Figure 1.5 illustrates this effect in a 28 dBm/26 dBm P_{SAT}/P_{1dB} with 40 % peak power-added efficiency (PAE). As can be seen, whenever the PA has to transmit a signal showing high PAPR, the average power of the signal is required to be well below the nonlinear region of the PA in order to avoid distortion. The nonlinear region is usually defined from the P_{1dB} point onwards. The difference between the P_{1dB} and the

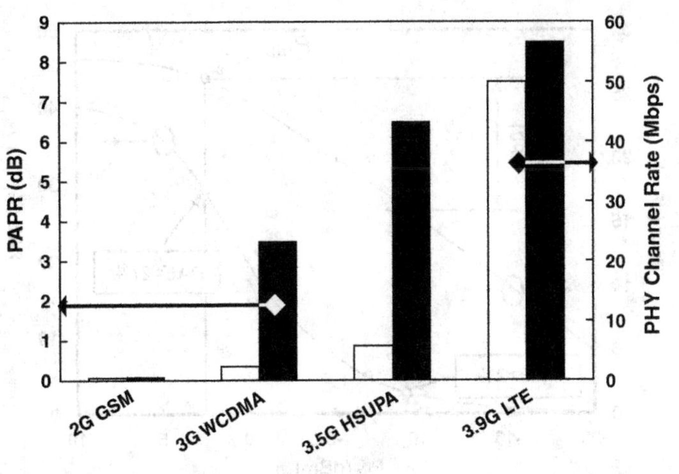

Fig. 1.4 Evolution of physical channel data rate and PAPR for different mobile communication standard generations: GSM, WCDMA, HSUPA and LTE

actual transmitted power is the power back-off, whose value depends on the linear characteristics of the PA and the type of modulation. However, what is clear is that the nonlinear region is the one showing the highest PA efficiency, whereas the efficiency of the PA drops as the power back-off increases. Following the performance of the PA in Fig. 1.5, if a power back-off of 10 dB from the P_{1dB} point was required the PAE would drop from 27 to 2 % and the channel average output power would be limited to 16 dBm. Consequently, a 16 dBm/50 Ω (~40 mW) output signal would require a power consumption of 2 W!

This again gives a clear idea of the need for high performance PAs because modern wireless standards come with spectrally efficient channels and a relatively large communication range, and because a non-optimized PA directly impacts the performance of the transceiver that is usually implemented in mobile, battery-powered devices.

CMOS Technology

The silicon-based CMOS technology can be regarded as the worst process for PA integration. However, CMOS processes have several major advantages: high availability and integrability, and very low cost. In fact, among the integrated circuit (IC) technologies, CMOS processes are known for their low cost mainly due to their widespread use in digital applications. CMOS processes account for 80 % of IC production so that whenever low cost and high production volumes are an issue, CMOS is unbeatable among the other IC technologies [15, 16].

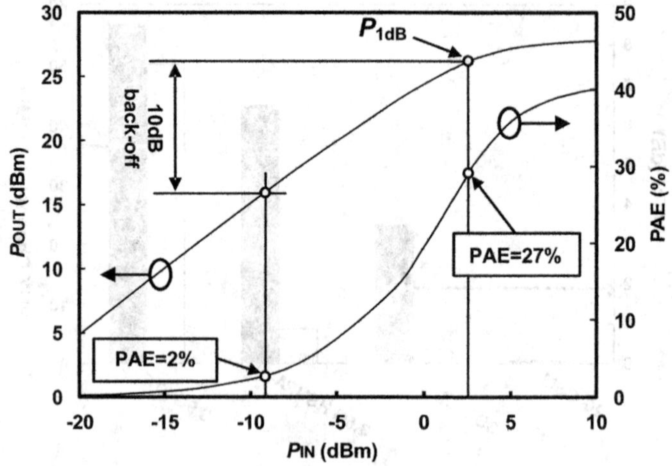

Fig. 1.5 Effect of the power back-off in a power amplifier

However, the CMOS technology also has several major drawbacks regarding high performance PA implementation. Although they are treated in more detail in Chap. 4, they are briefly outlined here.

The first major limitation is the low transistor breakdown voltage of submicrometer CMOS transistors along with the issue of hot carrier degradation, which leads to low supply voltage (V_{DD}) [17]. If the supply voltage must be low, for a specific output power level the current must be increased and consequently metal tracks and transistors must be wider. All this has an impact on the size of the chip and on poor efficiencies [18]. In addition to this, PAs dealing with high current levels and limited efficiencies must be concerned about power dissipation in the small area of the chips [19]. High temperatures in a chip degrade the performance of the PA. All these problems become even more critical as the CMOS processes scale down [19].

Another problem is that the CMOS transistor may also show high knee voltage (V_{KNEE}); this leads to even lower voltage headroom at the output and consequently even poorer efficiencies and output power levels.

Finally, if there is a need for large output power levels despite the low transistor breakdown voltage of transistors, then in these cases some form of impedance transformation is required. The impedance transformation suffers from high loss caused by the highly conductive substrate as well as the thin metal and dielectric layers [20].

However, all these disadvantages can be thought of as challenges that can be overcome if the PA design takes advantage of the process capabilities and smart techniques are applied. In fact, as CMOS technology is becoming a good choice for RF ICs, CMOS PAs have their chance due to the advantages of cost, integrability and size.

Organization of the Book

This book consists of eight chapters. It describes the fundamentals, theory, design and tests of linear CMOS Power amplifiers for RF applications.

The main metrics concerning PAs are found in Chap. 2 for linear CMOS PAs. AM–AM and AM-PM distortions and the parameters that quantify PA efficiency are part of this chapter. In a real situation, a PA will handle digitally modulated channels, and the corresponding standard sets limits for the quality of the output signal with specific parameters like the EVM or ACLR. Finally, although not widely used, the power capability is a parameter that allows the performance of different power amplifier classes to be compared in general.

The different classes of PAs are presented in Chap. 3. PAs are grouped depending on how the RF transistor works for the specific PA: as a current source or as a switch. The current source mode PAs fall in classes A to C, whereas classes D and E are for switch mode PAs. The class F PA is treated separately as it falls between current source and switch-type amplifiers.

Chapter 4 addresses the issues that a designer faces during the PA optimization in CMOS processes. As mentioned before, the main issue stems from the low breakdown voltage of CMOS transistors. However, other effects such as impedance transformation, limitations from the knee voltage, substrate losses, stability and transistor parasitics are also treated in detail.

Chapter 5 is devoted to PA architectures that are of interest for fully integrated PAs with relatively high output power levels or high efficiency. Power combination or the Doherty architecture as well as dynamic supply, digital predistortion techniques and the use of cascade transistors are interesting solutions in order to boost PA efficiencies or output power levels in CMOS processes with limited supply.

Chapter 6 deals with the design flow of linear CMOS PAs. It starts with the description of a PA model that helps with the first steps of PA design. The main parameters that determine PA performance are included in the model: PA biasing, current consumption, supply and breakdown voltages, inductor quality factors (Q) and power gain. The following steps of the PA's design flow, at the schematic and layout levels, are discussed next. Special attention is paid to the issues of integrated inductors within PAs and the extra considerations that a designer must know in order to optimize inductor performance.

The test setups that are required to characterize a PA are described in Chap. 7. The PA performance can be tested by means of single-tone measurements from which the P_{1dB}, the P_{SAT} or the PAE can be extracted. On the other hand, tests based on a digital channel allow the spectral regrowth or the ACLR and the EVM to be measured. Test setups for measuring inductors for PAs are also included.

Finally, Chap. 8 presents the conclusions of the book.

References

1. Zolfaghari A, Razavi B (2003) A low-power 2.4-GHz transmitter/receiver CMOS IC. IEEE J Solid-State Circ 38(2):176–183
2. Wei-Yi H, Jia-Wei L, Kuo-Chi T, Yong-Hsiang H, Chao-Liang C, Hung-Ta T, Yi-Shun S, Shao-Chueh H, Sao-Jie C (2010) A 0.18-μm CMOS RF transceiver with self-detection and calibration functions for bluetooth V2.1 + EDR applications. IEEE Trans Microw Theory Techn 58(5):1367–1374
3. Si W, Weber D, Abdollahi-Alibeik S, MeeLan L, Chang R, Dogan H, Gan H, Rajavi Y, Luschas S, Ozgur S, Husted P, Zargari M (2008) A single-chip cmos bluetooth v2.1 radio SoC. IEEE J Solid-State Circ 43(12):2896–2904
4. Ahola R, Aktas A, Wilson J, Kishore Rama R, Jonsson F, Hyyrylainen I, Brolin A, Hakala T, Friman A, Makiniemi T, Hanze J, Sanden M, Wallner D, Yuxin G, Lagerstam T, Noguer L, Knuuttila T, Olofsson P, Ismail M (2004) A single-chip CMOS transceiver for 802.11a/b/g wireless LANs. IEEE J Solid-State Circ 39(12):2250–2258
5. Pengfei Z, Der L, Dawei G, Sever I, Bourdi T, Lam C, Zolfaghari A, Chen J, Gambetta D, Baohong C, Gowder S, Hart S, Huynh L, Nguyen T, Razavi B (2005) A single-chip dual-band direct-conversion IEEE 802.11a/b/g WLAN transceiver in 0.18-μm CMOS. IEEE J Solid-State Circ 40(9):1932–1939
6. Nathawad L, Weber D, Shahram A, Chen P, Syed E, Kaczynski B, Alireza K, MeeLan L, Limotyrakis S, Onodera K, Vleugels K, Zargari M, Wooley B (2006) An IEEE 802.11a/b/g

SoC for embedded WLAN applications. In: IEEE international digest of technical papers solid-state circuits conference (ISSCC 2006), pp 1430–1439
7. Ingels M, Giannini V, Borremans J, Mandal G, Debaillie B, Van Wesemael P, Sano T, Yamamoto T, Hauspie D, Van Driessche J, Craninckx J (2010) A 5 mm^2 40 nm LP CMOS transceiver for a software-defined radio platform. IEEE J Solid-State Circ 45(12):2794–2806
8. Sowlati T, Agarwal B, Cho J, Obkircher T, El Said M, Vasa J, Ramachandran B, Kahrizi M, Dagher E, Wei-Hong C, Vadkerti M, Taskov G, Seckin U, Firouzkouhi H, Saeidi B, Akyol H, Yunyoung C, Mahjoob A, D''Souza S, Chieh-Yu H, Guss D, Shum D, Badillo D, Ron I, Ching D, Feng S, Yong H, Komaili J, Loke A, Pullela R, Pehlivanoglu E, Zarei H, Tadjpour S, Agahi D, Rozenblit D, Domino W, Williams G, Damavandi N, Wloczysiak S, Rajendra S, Paff A, Valencia T (2009) Single-chip multiband WCDMA/HSDPA/HSUPA/EGPRS transceiver with diversity receiver and 3G DigRF interface without SAW filters in transmitter/3G receiver paths. In: IEEE international digest of technical papers solid-state circuits conference (ISSCC 2009), pp 116–117,117a
9. Nilsson M, Mattisson S, Klemmer N, Anderson M, Arnborg T, Caputa P, Ek S, Lin F, Fredriksson H, Garrigues F, Geis H, Hagberg H, Hedestig J, Hu H, Kagan Y, Karlsson N, Kinzel H, Mattsson T, Mills T, Fenghao Ma rtensson A, Nicklasson L, Oredsson F, Ozdemir U, Park F, Pettersson T, Pahlsson T, Palsson M, Ramon S, Sandgren M, Sandrup P, Stenman A, Strandberg R, Sundstrom L, Tillman F, Tired T, Uppathil S, Walukas J, Westesson E, Xuhao Z, Andreani P (2011) A 9-band WCDMA/EDGE transceiver supporting HSPA evolution. In: IEEE international digest of technical papers solid-state circuits conference (ISSCC 2011), pp 366–368
10. Afsahi A, Behzad A, Magoon V, Larson LE (2010) Linearized dual-band power amplifiers with integrated baluns in 65 nm CMOS for a 2x2 802.11n MIMO WLAN SoC. IEEE J Solid-State Circ 45(5):955–966
11. Kanda K, Kawano Y, Sasaki T, Shirai N, Tamura T, Kawai S, Kudo M, Murakami T, Nakamoto H, Hasegawa N, Kano H, Shimazui N, Mineyama A, Oishi K, Shima M, Tamura N, Suzuki T, Mori T, Niratsuka K, Yamaura S (2012) A fully integrated triple-band CMOS power amplifier for WCDMA mobile handsets. In: IEEE international digest of technical papers solid-state circuits conference (ISSCC 2012), pp 86–88
12. Delaunay N, Deltimple N, Kerherve E, Belot D (2011) A RF transmitter linearized using cartesian feedback in CMOS 65 nm for UMTS standard. In: IEEE topical conference on power amplifiers for wireless and radio applications (PAWR 2011), pp 49–52
13. Ba SN, Khurram W, Zhou GT (2010) Efficient lookup table-based adaptive baseband predistortion architecture for memoryless nonlinearity. EURASIP J Adv Signal Process. doi:10.1155/2010/379249
14. Hildersley J (2012) Managing 4G handset RF power consumption, electronics design, strategy. News EDN Eur 57(11):16-21
15. Alvarado U, Bistué G, Adín I (2011) Low power RF circuit design in standard CMOS technology. Springer, Berlin
16. Reynaert P, Steyaert M (2006) RF power amplifiers for mobile communications. Springer, The Netherlands
17. Hella MM, Ismail M (2002) RF CMOS power amplifiers: theory, design and implementation. Kluwer Academic Publishers, Boston
18. Hastings A (2001) The art of analog layout. Prentice Hall, Upper Saddle River
19. Aoki I, Kee SD, Rutledge DB, Hajimiri A (2002) Fully integrated CMOS power amplifier design using the distributed active-transformer architecture. IEEE J Solid-State Circ 37(3):371–383
20. Aoki I, Kee SD, Rutledge DB, Hajimiri A (2002) Distributed active transformer-a new power-combining and impedance-transformation technique. IEEE Trans Microw Theory Techn 50(1):316–331

Chapter 2
Power Amplifier Fundamentals: Metrics

Abstract This chapter details the metrics that are usually used to evaluate the performance of a PA. PA metrics can be combined into two main groups: linearity and efficiency, and they can be related to either single-tone or two-tone or digital channel tests. In addition, this chapter discusses the effect of the 1 dB compression (P_{1dB}) point and saturated power (P_{SAT}) parameters in the digital channel distortion. Finally, stability and power capability metrics are also included.

AM–AM Distortion

AM–AM distortion refers to the amplitude distortion of any amplifier that is driven in a nonlinear condition. This distortion is reflected in the compression effect that can be observed in the P_{IN}–P_{OUT} curve of a PA when a single-tone test is performed or the intermodulation products in the case of a two-tone test.

Saturated Power and One dB Compression Point

Figure 2.1 illustrates the P_{IN}–P_{OUT} curve of a PA in dBm along with the P_{1dB} and P_{SAT} parameters. These two parameters are a classic way of describing the performance of a PA in terms of linearity. The P_{1dB} is generally given as an output power value and refers to the output power level at which the power gain drops 1 dB from the linear value. The P_{1dB} value represents a practical limit between the linear and the nonlinear region of a PA. However, the 1 dB compression point actually represents a moderate rather than a weak nonlinear point [1]. In practice the PA output power levels considered acceptable for many communication standards are well below the P_{1dB}.

The P_{SAT} parameter indicates the maximum output power that a PA can achieve. When working at this power level, the PA is extremely nonlinear with a

H. Solar Ruiz and R. Berenguer Pérez, *Linear CMOS RF Power Amplifiers*,
DOI: 10.1007/978-1-4614-8657-2_2,

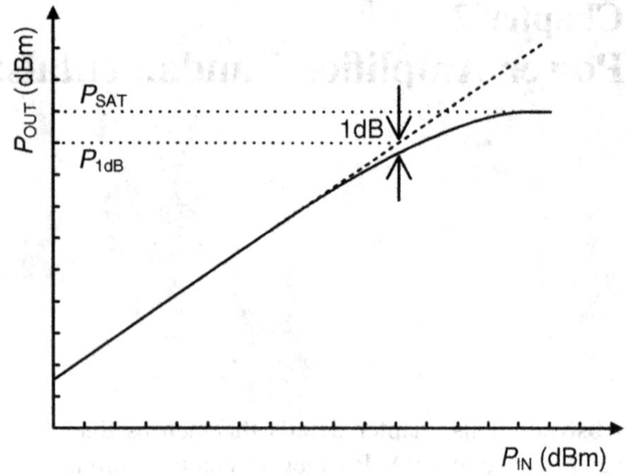

Fig. 2.1 Typical AM–AM distortion in the P_{IN}–P_{OUT} curve of a linear PA

power gain value that is much lower than the linear gain at low power levels. As the P_{SAT} is in the strong nonlinear region, the P_{SAT} output power level is clearly unacceptable for the channels transmitted. However, the distance between the P_{SAT} and the P_{1dB} gives interesting information about the practical performance of the PA, as we will discuss in the next sections.

Third Order Intercept Point

If two carriers or tones, usually close in frequency, are applied to the input, the nonlinearities of the PA are reflected in the appearance of several distortion products of different orders. The ones that are of interest due to their possible detrimental effects are the intermodulation products (IM_n), which are odd order products close to the fundamental carrier. The third order products (IM_3) are placed at frequencies $2\omega_2-\omega_1$ and $2\omega_1-\omega_2$. Higher order products such as the fifth order intermodulation products (IM_5) at frequencies $3\omega_2-2\omega_1$ and $3\omega_1-2\omega_2$ have smaller amplitudes and are more distant from the fundamental carrier. These higher order contributions can then be ignored when the PA operates well below the nonlinear region of the PA, but they are dominant contributors when the PA works in the compression regime. On top of that they contribute to the third order intermodulation products. A typical spectrum is shown in Fig. 2.2. The intermodulation products appear at either side of each carrier with frequency spacing equal to that of the two input carriers.

As shown in Fig. 2.3, the third order intercept point (IP3) represents the intersection between the extrapolated P_{IN}–P_{OUT} curve of the fundamental tones at the output, which has a 1:1 slope and the extrapolated P_{IN}–P_{OUT} curve of the third order IM products with a 3:1 slope. The IP3 is quantified as a power level either at the input (IIP3) or at the output (OIP3).

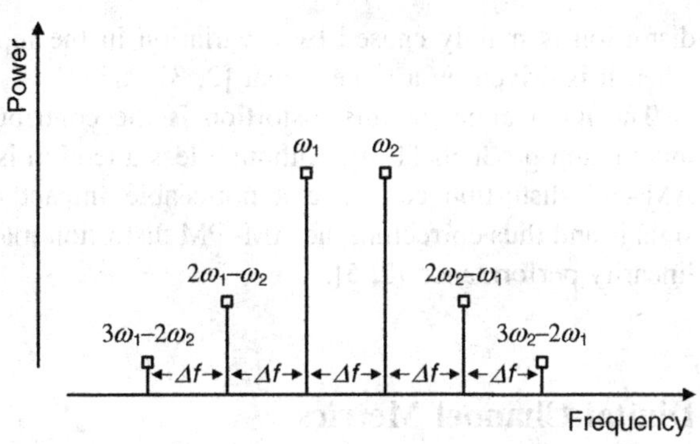

Fig. 2.2 Intermodulation products in a two-tone test of an amplifier

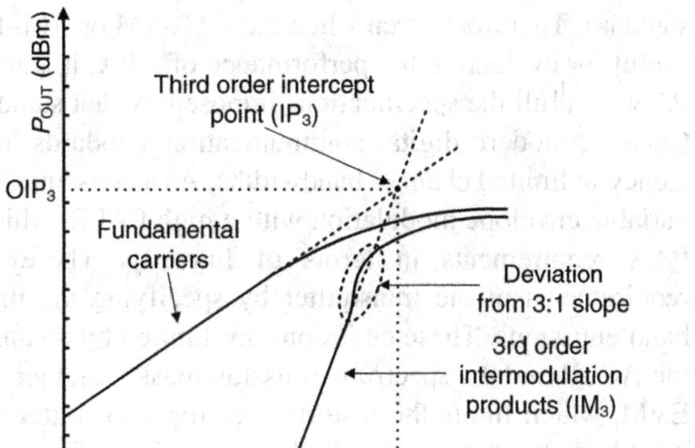

Fig. 2.3 Single carrier and IM_3, P_{IN}–P_{OUT} curves and IP_3 calculation

To correctly evaluate the IP3, the extrapolation of the curves should be done well inside the linear region of the amplifier. At higher output power levels the actual IM3 curve of the amplifier deviates noticeably from the 3:1 slope as a consequence of high order products affecting the theoretical 3:1 line. This deviation from the theoretical slope in the high power levels of the PA (the ones of most interest) makes the IP3 point a less widespread parameter for evaluating the linearity performance of a PA. However, this type of distortion is of the same nature as the so-called spectral regrowth, which is of importance in communication standards and will be treated in a later section.

AM–PM Distortion

AM–PM distortion refers to the change in the phase difference between the output and input signals as a function of the input signal level. Fig. 2.4 shows the typical behavior of AM–AM and AM–PM distortion in a linear PA. The AM–PM distortion is mainly caused by a variation in the input impedance of the device when it is driven by a large signal [2, 3].

The main effect of this distortion is the contribution to asymmetrical intermodulation products [1, 4]. Although less attention is paid to AM–PM distortion, AM–PM distortion can have a noticeable impact on phase modulated digital signals and thus correcting the AM–PM distortion may provide the PA with better linearity performance [2, 5].

Digital Channel Metrics

PAs are designed to be the final stage of a transmitter for a specific communication standard. Therefore, even when the AM–AM or AM–PM distortion parameters are useful for evaluating the performance of a PA, it is more important to know if the PA will fulfill the specifications imposed by that standard. As we mentioned in the Chap. 1, modern digital communication standards look to increase spectral efficiency in limited channel bandwidths. As a consequence, the digital channel shows variable envelope modulation with a high PAPR, which has a direct impact on the PA's requirements in terms of linearity. These standards set the linearity requirements of the transmitter by specifying the maximum permissible out-of-band emissions. These emissions are limited by means of two main specifications: the ACLR and the spectrum emission mask. Another important specification is the EVM, which limits the distortion of the transmitter in the modulated channel. It must be noted that the standard specification refers to the transmitter and therefore the values should be shared with all the components of the transmission chain.

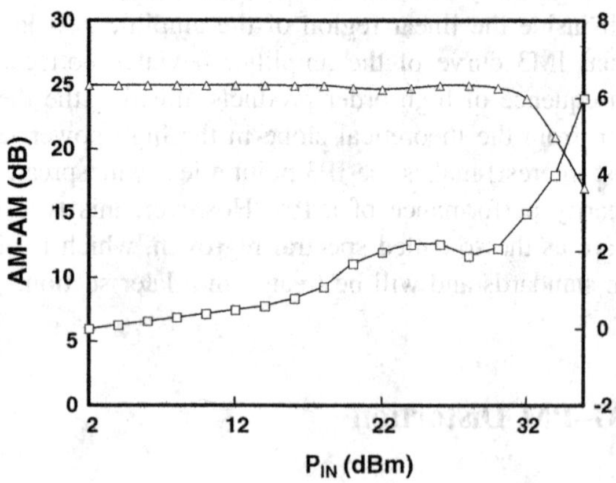

Fig. 2.4 Example of AM–AM and AM–PM distortion in a class AB PA

Spectral Regrowth

A similar effect that is observed in a two-tone test due to the nonlinearities of the PA appears if the input signal is a modulated channel. In this case the nonlinearities of the PA generates an intermodulation band that stretches out to three times the original band limit of the channel in the case of the third order distortion products or five times these limits for fifth order intermodulation. This stretching out effect is called spectral regrowth, and its limits are regulated by communication standards. Figure 2.5 shows an example of the spectral regrowth caused by the nonlinearities of a PA in a 40 MHz 802.11n channel. The spectra of the normalized channel at the input and the output of the PA can be observed.

Adjacent Channel Leakage Ratio

The spectral regrowth of most concern in a channelized communication standard is the one caused by the third order intermodulation distortion because it lies closest to the main signal. Standards usually limit this spectral regrowth by specifying the adjacent channel leakage ratio (ACLR). The ACLR refers to the amount of power a device transmitting in a channel leaks into its adjacent channel, as seen in Fig. 2.6.

Take as an example the 3GPP LTE mobile standard, which sets a maximum limit to the power ratio of 30 dBc for an adjacent LTE channel into the wanted LTE channel with an output power of 23 dBm for mobile transmitters [6]. These values serve as an initial set for specifying the linearity requirements of a PA intended for LTE applications.

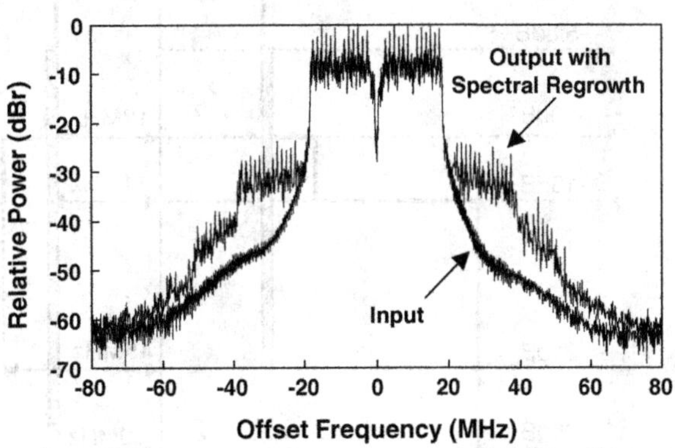

Fig. 2.5 Spectral regrowth in an 802.11n 40 MHz channel

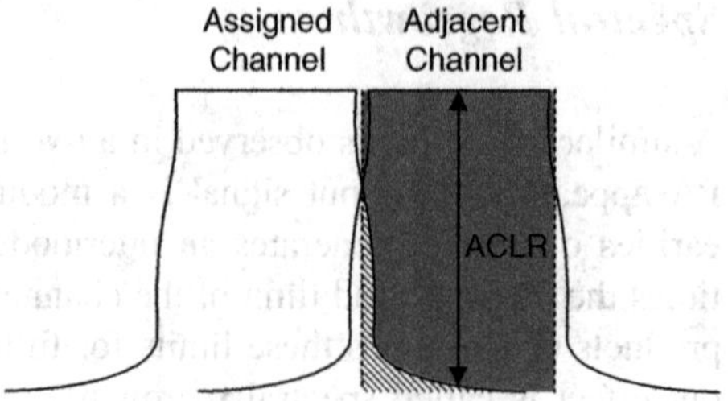

Fig. 2.6 Adjacent channel leakage ratio

Spectrum Emission Mask

In addition to the above ACLR specification, communication standards also set a mask of power levels relative to the power at the center of the channel, and these limits cannot be exceeded. As the nonlinearities of PAs cause spectral regrowth, the spectrum emission mask is another specification that must be taken into account when fixing the required linearity of a PA or the maximum available output power for a given PA. Figure 2.7 shows the spectrum emission mask of a 20 MHz channel for the 3GPP LTE standard [6]. It must be noted that the spectrum mask limits are always given for a specific resolution bandwidth, and they must be taken into consideration during testing. In this case, the resolution bandwidth is set to 1 MHz, except for the closest frequency offset (10–11 MHz), which is 30 kHz. Figure 2.7 shows the spectrum emission mask normalized to 1 MHz resolution bandwidth.

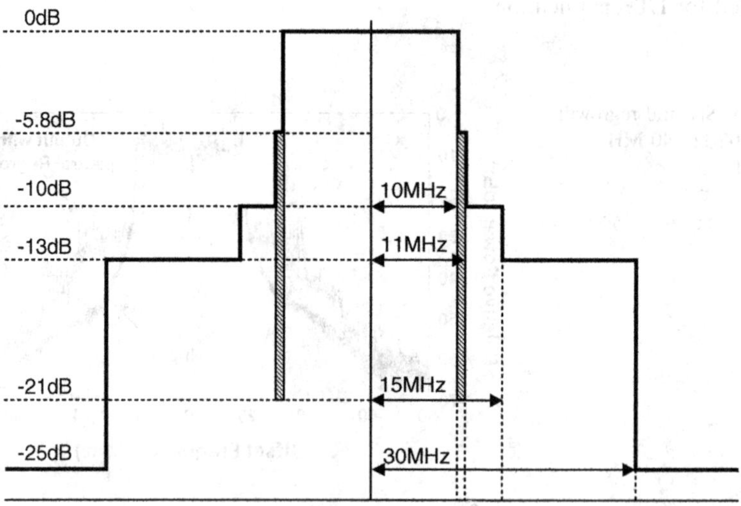

Fig. 2.7 Spectrum emission mask of a 20 MHz LTE channel

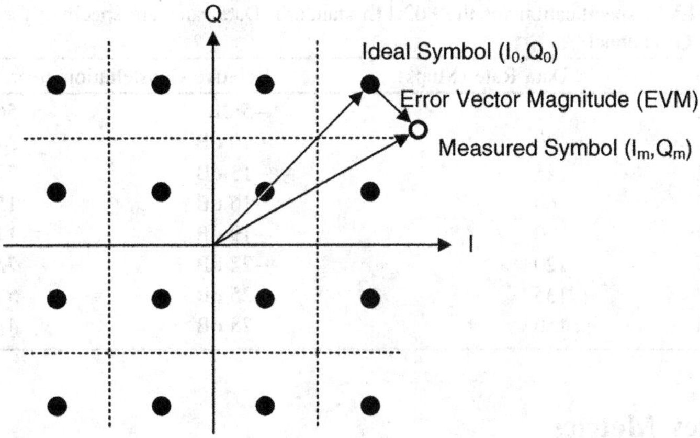

Fig. 2.8 Error vector magnitude concept in a 16QAM constellation

Error Vector Magnitude

Another linearity metric that can be usually found as a specification in communication standards is the error vector magnitude (EVM). The error vector quantifies the transmit modulation accuracy, as shown in Fig. 2.8, and it is defined as the difference between the ideal constellation point and the actual transmitted constellation point. The distortion and nonlinearities of the transmitter causes the actual constellation points to deviate from the ideal locations. Specifically, the nonlinearities of the PA cause the EVM to degrade so that either the output power must be reduced or the PA linearity improved in order to fulfill this specification.

Normally the EVM is quantified as the average rms value of a number of symbols for the modulated signal and for a specific transmitted sequence. Its value is provided by the standards in decibels or as a percentage, and the standard tends to impose more stringent values as the number of constellation points in the modulation scheme increases. The value of the EVM in decibels can be easily related to the value as a percentage by means of (2.1).

$$EVM(dB) = 20\log\left(\frac{EVM(\%)}{100}\right) \tag{2.1}$$

An interesting example of specified EVM values are the ones found in the 802.11n standard because it allows four different modulation schemes: BPSK, QPSK, 16QAM and 64QAM [7]. Their values are shown in Table 2.1.

Table 2.1 EVM specification for the 802.11n standard. Data rates are specified for a 40 MHz and 400 ns GI channel

Modulation	Data Rate (Mbps)	Relative Constellation error	
BPSK	15	−5 dB	56.2 %
QPSK	30	−10 dB	31.6 %
QPSK	45	−13 dB	22.4 %
16QAM	60	−16 dB	15.8 %
16QAM	90	−19 dB	11.2 %
64QAM	120	−22 dB	7.9 %
64QAM	135	−25 dB	5.6 %
64QAM	150	−28 dB	4 %

Efficiency Metrics

PA efficiency parameters quantify the amount of consumed DC power that is turned into RF power. Efficiency is a crucial parameter for a PA. As mentioned in Chap. 1, the power consumption of the PA is a significant amount of the total power of the transceiver. Therefore, the efficiency of a PA will directly impact the efficiency of the whole transceiver and the battery lifetime if the PA is used in portable applications. Moreover, the quality of a PA in terms of efficiency can also be an issue for heat dissipation in high power level or integrated PAs. Two main metrics are commonly employed for evaluating the efficiency of RF PAs: drain efficiency (η_d) and power-added efficiency (PAE).

Drain Efficiency

Drain efficiency is the ratio of the RF power at the output of the PA (P_{OUT}) and the DC power consumption (P_{DC}), as expressed in (2.2).

$$\eta_d = 100 \times \frac{P_{OUT}}{P_{DC}} \tag{2.2}$$

Power-Added Efficiency

PAE is defined as the ratio between the RF power that the PA adds to the input power and the DC power consumed in order to get this addition. It is mathematically expressed in (2.3).

$$PAE = 100 \times \frac{P_{OUT} - P_{IN}}{P_{DC}} \tag{2.3}$$

PAE is the usual metric to evaluate efficiency because it is a more complete metric of a PA efficiency performance. As (2.4) shows, the PAE takes into account the effect of the power gain (G) and thus penalizes PA designs with poor gain. Therefore, the PAE also considers the requirements of the previous PA driver and punishes a poor gain PA that requires the driver to deal with higher power levels.

$$PAE = 100 \times \frac{P_{OUT} - \frac{P_{OUT}}{G}}{P_{DC}} = 100 \times \frac{P_{OUT}}{P_{DC}}\left(1 - \frac{1}{G}\right) \tag{2.4}$$

The gain correction factor that the PAE adds to the drain efficiency is graphically represented in Fig. 2.9 for a 25 dB gain linear PA. It can be observed that while a PA is in its linear region, the PAE and the drain efficiency curves stay close one to another, but as the gain is reduced due to compression the PAE decreases and separates from the η_d.

Finally, as Fig. 2.10 illustrates, it must be noted that the P_{DC} does not necessarily have to be constant as the P_{IN} increases. In fact, for a class AB linear PA, the P_{DC} shows a noticeable increase whenever the PA enters compression due to the clipping effects of the output current. In class AB PAs, the actual current consumption of the PA separates from the quiescent current in the nonlinear region of the PA and the P_{DC} calculation must be made following (2.5).

$$P_{DC} = V_{DC} \times \frac{1}{T}\int_{0}^{T} i_{DS}(t) \cdot dt \tag{2.5}$$

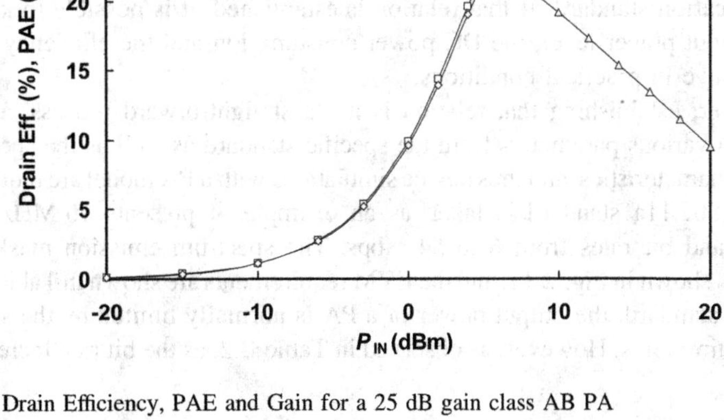

Fig. 2.9 Drain Efficiency, PAE and Gain for a 25 dB gain class AB PA

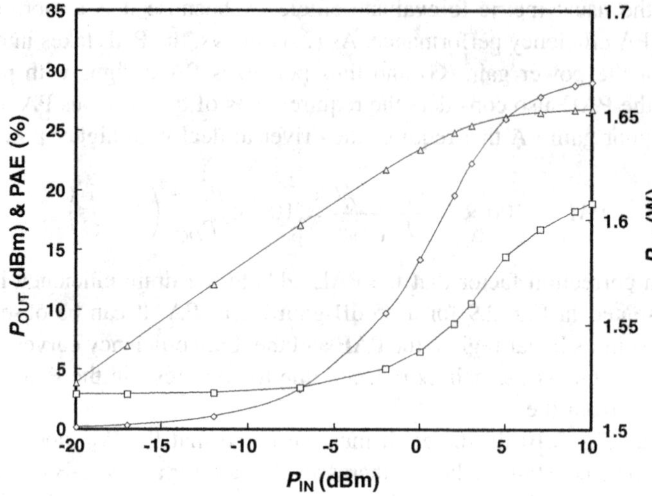

Fig. 2.10 Typical P_{OUT}, PAE and P_{DC} curves vs P_{IN} for a 27 dBm P_{SAT} class AB PA

The increment of the P_{DC} in linear class AB PAs must be considered in the PAE calculation and will be discussed in more detail in Chap. 3 and Chap. 6.

Power Back-Off

Since P_{1dB} and P_{SAT} are the usual metrics for evaluating the linearity performance of a PA, it is of great interest to know how to relate them to the maximum output power a PA may transmit for a specific communication standard. In complex digital channels with a high PAPR an important parameter is the power back-off, i.e. the amount of the necessary power reduction from a P_{1dB} or P_{SAT} in order to be compliant with the EVM and spectrum emission mask requirements imposed by a communication standard. If that relation is established, it is possible to know the actual output power level, the DC power consumption and the efficiency that the PA will have in practical conditions.

However, establishing that relation is not a straightforward process, as it will depend on various parameters from the specific standard as well as the specific PA linearity characteristics, and thus device simulations with a PA model are required [8].

If the 802.11a standard is taken as an example, it presents 16 MHz OFDM channels and bit rates from 6 to 54 Mbps. The spectrum emission mask of this standard is shown in Fig. 2.11, and the EVM requirements are shown in Table 2.2 [9].

In this standard, the output power of a PA is normally limited by the spectrum mask requirements. However, as observed in Table 2.2, as the bit rate increases the

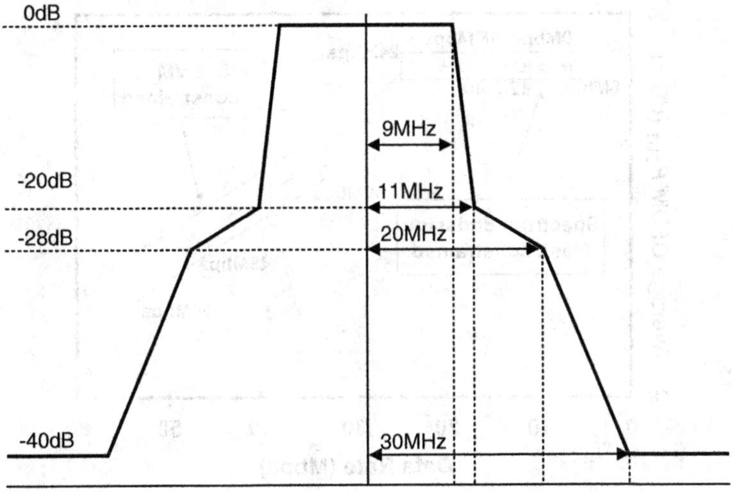

Fig. 2.11 Spectrum emission mask of a 16 MHz 802.11a channel (not to scale)

Table 2.2 EVM specification of the 802.11a standard

Modulation	Data Rate (Mbps)	Relative Constellation error (dB)
BPSK	6	−5
QPSK	9	−8
QPSK	12	−10
16QAM	18	−13
16QAM	24	−16
64QAM	36	−19
64QAM	48	−22
64QAM	54	−25

EVM requirements are the limitation of the output power because the EVM requirements are more stringent for the highest bit rates.

The maximum output power levels for each bit rate will then look like Fig. 2.12, where the output power is constrained by the spectrum emission mask for the lowest bit rates, but for the highest bit rates the EVM values are the main limitation. Therefore, the power back-off from either the P_{SAT} or the P_{1dB} parameters must increase with increases in bit rate.

These previous considerations can be quantified by means of several simulations using a PA in which the P_{1dB}, the P_{SAT} and the saturation level (the amount of gain compression at the saturation point) parameters were changed [10]. An 802.11a OFDM channel was used and the distortions caused by the nonlinearity of the PA were measured at the output. Figs. 2.13 and 2.14 show the results of these simulations. The P_{SAT} and the P_{1dB} parameters were swept while checking the fulfillment of the spectrum emission mask and the 54 Mbps EVM requirements of

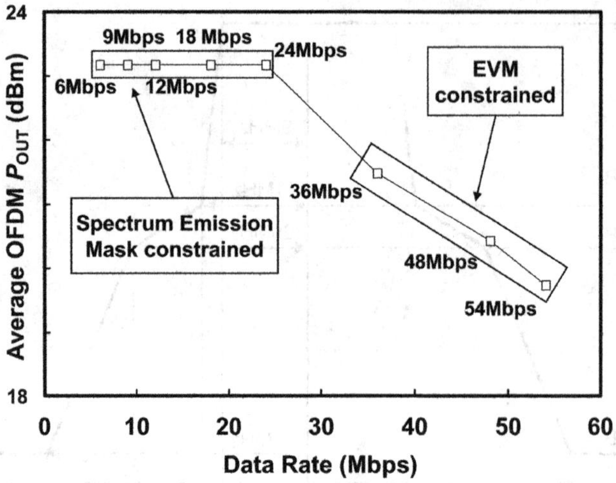

Fig. 2.12 Maximum output power levels vs. data rate for the 802.11a standard

the 802.11a standard since 54 Mbps is the bottleneck in terms of output power. In Fig. 2.13 the P_{1dB} is fixed (23 dBm) whereas the P_{SAT} varies (26–29 dBm), and in Fig. 2.14 the P_{1dB} is varied (21–24 dBm), while the P_{SAT} is kept constant (27 dBm). As Fig. 2.13 shows, in order to fulfill the standard at 54 Mbps, i.e. the −25 dB EVM specification, a 4 dB power back-off from the P_{1dB} is required. It is also worth noting that the P_{SAT} value does not affect the result. Given that the EVM at 54 Mbps is quite stringent, the PA has to work in very linear conditions so that the actual output power is sufficiently far from P_{SAT}. In such a case, very few signal peaks enter high compression and P_{SAT} variations do not affect the output power level that complies with the 54 Mbps EVM requirements. Therefore, in order to comply with the EVM at 54 Mbps, the P_{1dB} is the reference value and

Fig. 2.13 EVM and Spectrum Emission Mask compliant 54 Mbps OFDM average output power levels vs. P_{SAT} variations with P_{1dB} fixed value of 23 dBm

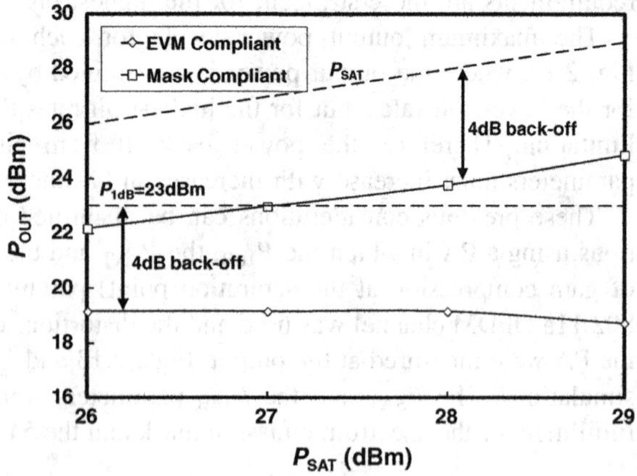

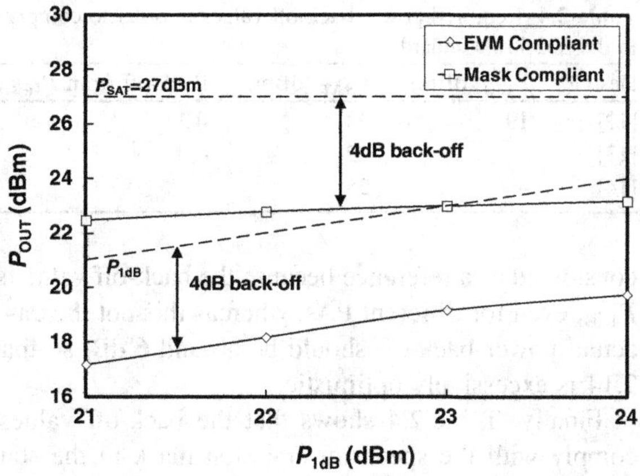

Fig. 2.14 EVM and Spectrum Emission Mask compliant 54 Mbps OFDM average output power levels vs. P_{1dB} variations with P_{SAT} fixed value of 27 dBm

optimizing the P_{1dB} would result in the highest power levels at 54 Mbps. This also is confirmed by the simulations presented in Fig. 2.14. In this case, where the P_{1dB} is varied while keeping the P_{SAT} constant, the output power level at 54 Mbps is almost linearly affected.

However, as Fig. 2.13 shows, if the spectrum emission mask is the only requirement, the power back-off would only be 4 dB from the P_{SAT}, so it is possible work in moderate nonlinear conditions of the PA. In this case, by being so close to the P_{SAT}, variations in this parameter linearly affect the output power which complies with the spectrum emission mask and, as shown in Fig. 2.14, even though the P_{1dB} is varied, the output power level is not affected in practice.

Therefore, in order to design a PA with high OFDM output power values in the 802.11a standard, it is necessary to first maximize the P_{1dB} parameter. This is because the bottleneck of this standard in terms of output power is the one at 54 Mbps due to the stringent specification of the EVM.

The absolute output power values in Figs. 2.13 and 2.14 must be carefully considered because the model does not include all the nonlinear effects of PAs (such as AM–PM). However, in taking a look at the state-of-the-art, those values can be corrected. As Table 2.3 shows, some reports on 5 GHz WLAN power amplifiers [2, 11] and transceivers [12, 13] give some information about the required power back-off values in order to comply with the EVM requirements at 54 Mbps. Table 2.3 confirms that the P_{1dB} is the parameter that must be

Table 2.3 Required power back-off values in order to comply with the EVM at 54 Mbps in the 802.11a standard

Design	P_{1dB} (dBm)	P_{SAT} (dBm)	Back-off from P_{1dB} (dB)	Back-off from P_{SAT} (dB)
[2]	25.4	26.5	6	7.1
[11]	20.5	22	6	7.5
[12]	19	23	6.2	10.2
[13]	–	22	–	9.5

Table 2.4 Required power back-off values in order to comply with the spectrum emission mask in the 802.11a standard

Design	P_{1dB} (dBm)	P_{SAT} (dBm)	Back-off from P_{1dB} (dB)	Back-off from P_{SAT} (dB)
[12]	19	23	0.3	4.3
[13]	–	22	–	4.2
[14]	–	25	–	4.3

considered as a reference because the back-off value is kept constant relative to the P_{1dB}, even for different PAs, whereas this not the case for the P_{SAT}. However, the actual power back-off should be around 6 dB, so that the 4 dB in Figs. 2.13 and 2.14 is excessively optimistic.

Finally, Table 2.4 shows that the back-off values from the P_{SAT} in order to comply with the spectrum emission mask in the state-of-the-art are close to the previous results obtained from simulations. Table 2.4 also indicates the P_{SAT} should be the reference parameter in this case.

Transmit Power Levels

Continuing with the example of the 802.11a standard, with respect to the allowed transmit power levels, this standard refers to the FCC (Federal Communications Commission) regulation for unlicensed RF devices [15]. The Commission sets different power levels for each sub-band, which are shown in Fig. 2.15 along with European regulations [16].

For the lower sub-band a maximum power level of only 16 dBm is allowed, but the middle sub-band permits a maximum power level of 23 dBm and the upper

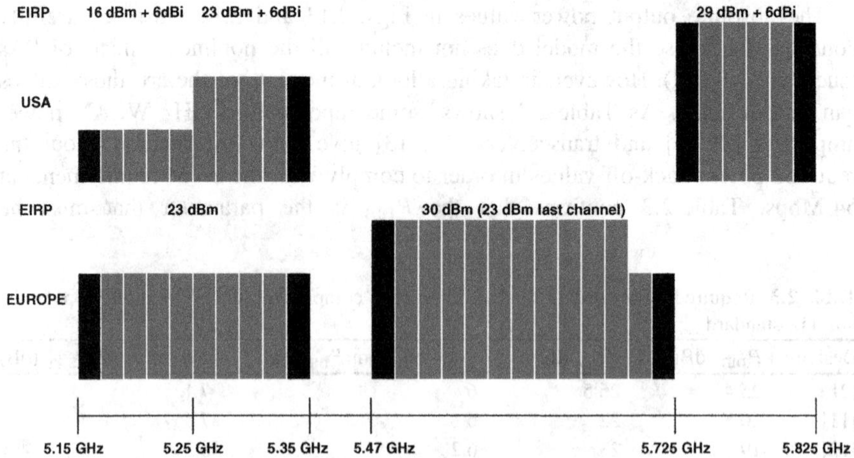

Fig. 2.15 Frequency plan and transmit power levels for the 5 GHz WLAN

sub-band maximum power level can be as high as 29 dBm. All the sub-bands can add a 6 dB gain antenna to increase the effective isotropic radiated power (EIRP) of the transmitter and hence the coverage.

If the objective is to achieve a minimum value of 16 dBm (the maximum OFDM output power for the lower sub-band) at all the data rates, then in accordance with the above conclusions the P_{1dB} should be at least 22 dBm to guarantee the EVM requirement at the 54 Mbps data rate.

As the P_{SAT} parameter does not influence the output power levels required to comply with the EVM requirement at the 54 Mbps data rate, its value can be chosen with greater flexibility. However, its value could be chosen so that the maximum output power level (23 dBm) of the middle sub-band is obtained while fulfilling the spectrum emission mask. Results from Table 2.4 indicate that a value of at least 27 dBm for P_{SAT} would be required for that purpose. Consequently, with a P_{1dB} of 22 dBm and a P_{SAT} of 27 dBm, the PA could work at 16 dBm for all the data rates in the lower sub-band and at 23 dBm for the spectrum emission mask limited data rates in the middle sub-band.

Stability

The large signal levels, transistor parasitics, magnetic coupling or PA nonlinearities may push the PA into an unstable region and make it oscillate. A practical and common parameter used to determine the stability of an amplifier is the Rollett stability factor or stability factor K [17–23]. Stability factor K ensures unconditional stability criteria based on the S-parameters of an amplifier modeled as a 2-port network, no matter what passive matching is placed at its input or output [1]. The conditions that must be fulfilled are shown in (2.6) and (2.7).

$$k = \frac{1 - |S_{11}|^2 - |S_{22}|^2 + |\Delta|^2}{2|S_{12}S_{21}|} \Rightarrow k > 1 \tag{2.6}$$

$$\Delta = S_{11}S_{22} - S_{12}S_{21} \tag{2.7}$$

It must be noted that the stability analysis based on stability factor K has some limitations. First, it assumes the linear behavior of the PA, whereas the PA may work in nonlinear conditions. However, it is possible to get around this issue if the stability analysis is performed based on large-signal S-parameter results either in simulations or measurements. The distinctive feature of large-signal S-parameters is that they are not only a function of frequency, but they also depend on the output power levels. Therefore, it is possible to perform a stability analysis of the PA with stability factor K at different power levels.

Another restriction is that the stability factor analysis, as it is presented, does not ensure the stability for multistage PAs [1, 24, 25]. Therefore, although a multistage amplifier considered as a single two-port can be analyzed with stability factor K some other analysis, such as a transient analysis, must be considered in the PA design process.

Power Capability

Power capability is a metric that is not usually evaluated in practical PA implementation; rather it serves to theoretically compare the relative stress that the different PA classes place on the active transistor. Since the power capability metric will be used in the next Chap. 3 when discussing the PA classes, it is presented in (2.8).

$$P_N = \frac{P_{OUT}}{v_{DS,peak} \cdot i_{DS,peak}} \tag{2.8}$$

In (2.8) $v_{\mathrm{DS,peak}}$ and $i_{\mathrm{DS,peak}}$ refer the drain voltage and current peak values, so P_{N} quantifies the amplitude of the output RF voltage and current values in order to reach a specific P_{OUT}.

Conclusions

This chapter has presented the parameters that must be considered in order to quantify the performance of an RF PA. Modern communication standards look to increase the spectral efficiency of the channels in limited channel bandwidths, so these channels show a high PAPR. Therefore, although single-tone tests are always important in order to obtain $P_{1\mathrm{dB}}$ or P_{SAT} and to allow different PAs to be compared, a PA is also required to verify its performance against realistic digital channels. Testing the PA with realistic channels allows for the determination of the actual efficiency and the power levels that the PA can handle for the specific standard while complying with requirements like the ACLR, the spectrum emission mask or the EVM.

References

1. Cripps SC (1999) RF power amplifiers for wireless communications. Artech House, Norwood
2. Elmala M, Paramesh J, Soumyanath K (2006) A 90-nm CMOS Doherty power amplifier with minimum AM–PM distortion. IEEE J Solid-State Circuits 41(6):1323–1332
3. Onizuka K, Ishihara H, Hosoya M, Saigusa S, Watanabe O, Otaka S (2012) A 1.9 GHz CMOS power amplifier With embedded linearizer to compensate AM–PM distortion. IEEE J Solid-State Circuits 47(8):1820–1827

4. Chowdhury D, Hull CD, Degani OB, Yanjie W, Niknejad AM (2009) A fully integrated dual-mode highly linear 2.4 GHz CMOS power amplifier for 4G WiMax applications. IEEE J Solid-State Circuits 44(12):3393–3402
5. Palaskas Y, Taylor SS, Pellerano S, Rippke I, Bishop R, Ravi A, Lakdawala H, Soumyanath K (2006) A 5-GHz 20-dBm power amplifier with digitally assisted AM–PM correction in a 90-nm CMOS process. IEEE J Solid-State Circuits 41(8):1757–1763
6. ETSI TS 136 101 V10.7.0 (2012-07). LTE; Evolved Universal Terrestrial Radio Access (E-UTRA); User Equipment (UE) radio transmission and reception (3GPP TS 36.101 version 10.7.0 Release 10). http://www.etsi.org. Accessed 1 April 2013
7. IEEE Std 802.11n-2009. Part 11: Wireless LAN Medium Access Control (MAC) and Physical Layer (PHY) Specifications-Amendment 5: Enhancements for Higher Throughput. http://standards.ieee.org. Accessed 1 April 2013
8. McFarland B, Shor A, Tabatabaei A (2002) A 2.4 & 5 GHz dual band 802.11 WLAN supporting data rates to 108 Mb/s. In: IEEE Tech Digest of Gallium Arsenide Integrated Circuit Symposium (GaAs IC 2002), Sunnyvale, pp 11–14
9. IEEE Std 802.11a-1999. Part 11: Wireless LAN Medium Access Control (MAC) and Physical Layer (PHY) Specifications-High-Speed Physical Layer in the 5 GHz Band, IEEE, 2000. http://standards.ieee.org. Accessed 1 April 2013
10. Software ADS of Agilent EEsof EDA. http://www.home.agilent.com. Accessed 1 April 2013
11. Yongwang D, Harjani R (2005) A high-efficiency CMOS +22-dBm linear power amplifier. IEEE J Solid-State Circuits 40(9):1895–1900
12. Behzad AR, Zhong Ming S, Anand SB, Li L, Carter KA, Kappes MS, Tsung-Hsien L, Nguyen T, Yuan D, Wu S, Wong YC, Victor F, Rofougaran A (2003) A 5-GHz direct-conversion CMOS transceiver utilizing automatic frequency control for the IEEE 802.11a wireless LAN standard. IEEE J Solid-State Circuits 38(12):2209–2220
13. Zargari M, Su DK, Yue CP, Rabii S, Weber D, Kaczynski BJ, Mehta SS, Singh K, Mendis S, Wooley BA (2002) A 5-GHz CMOS transceiver for IEEE 802.11a wireless LAN systems. IEEE J Solid-State Circuits 37(12):1688–1694
14. Kidwai AA, Nazimov A, Eilat Y, Degani O (2009) Fully integrated 23 dBm transmit chain with on-chip power amplifier and balun for 802.11a application in standard 45 nm CMOS process. In: IEEE Radio Frequency Integrated Circuits Symposium (RFIC 2009), Hillsboro, pp 273–276
15. FCC (Federal Communications Commission) Title 47 CFR (Code of Federal Regulations) Part 15: Radio Frequency Devices. Subpart E—Unlicensed National Information Infrastructure (U-NII) Devices. http://www.fcc. Accessed 1 April 2013
16. ECC/DEC/(04)08. ECC Decision of 12 Nov 2004 on the harmonized use of the 5 GHz frequency bands for the implementation of Wireless Access Systems including Radio Local Area Networks (WAS/RLANs) (ECC/DEC/(04)08). http://www.cept.org/ecc. Accessed 1 April 2013
17. Liu JY, Tang A, Ning-Yi Wang, Gu QJ, Berenguer R, Hsieh-Hung Hsieh, Po-Yi Wu, Chewnpu Jou, Chang M-F (2011) A V-band self-healing power amplifier with adaptive feedback bias control in 65 nm CMOS. In: IEEE radio frequency integrated circuits symposium (RFIC 2011), pp 1–4
18. Pornpromlikit S, Jeong J, Presti CD, Scuderi A, Asbeck PM (2010) A Watt-level stacked-FET linear power amplifier in silicon-on-insulator CMOS. IEEE Trans Microw Theory Tech 58(1):57–64
19. Xu Z, Gu QJ, Chang MCF (2011) A 100–117 GHz W-Band CMOS power amplifier With on-chip adaptive biasing. IEEE Microw Wireless Compon Lett 21(10):547–549
20. LaRocca T, Liu JYC, Chang MCF (2009) 60 GHz CMOS amplifiers using transformer-coupling and artificial dielectric differential transmission lines for compact design. IEEE J Solid-State Circuits 44(5):1425–1435
21. Komijani A, Natarajan A, Hajimiri A (2005) A 24-GHz, +14.5-dBm fully integrated power amplifier in 0.18 μm CMOS. IEEE J Solid-State Circuits 40(9):1901–1908

22. Chowdhury D, Reynaert P, Niknejad AM (2008) A 60 GHz 1 V + 12.3 dBm Transformer-Coupled Wideband PA in 90 nm CMOS. In: IEEE international digest of technical papers solid-state circuits Conference (ISSCC 2008), pp 560–635
23. Asada H, Matsushita K, Bunsen K, Okada K, Matsuzawa A (2011) A 60 GHz CMOS power amplifier using capacitive cross-coupling neutralization with 16 % PAE. In: European microwave integrated circuit conference (EuMIC 2011), pp 554–557
24. Jackson RW (2006) Rollett proviso in the stability of linear microwave circuits-a tutorial. IEEE Trans Microw Theory Tech 54(3):993–1000
25. Narhi T,Valtonen M (1997) Stability envelope-new tool for generalised stability analysis. In: IEEE MTT-S international microwave symposium digest (IMS 1997), pp 623–626

Chapter 3
Power Amplifier Fundamentals: Classes

Abstract This chapter analyzes the different PA operation modes. The different PA classes are traditionally classified into two main groups: current source amplifiers, which comprises classes A to C, and switch-type amplifiers, which make up classes E and D. The class F PA is treated separately as it falls between current source and switch-type amplifiers. The following sections describe each operation mode in detail.

Current Source PAs

This first group of amplifiers comprises the traditional class A, AB, B and C operation modes, where the transistor is considered as a current source. The main difference between these four classes relates to the bias voltage at the gate that modifies the current conduction angle. The linearity of these PAs falls off as the conduction angle decreases from class A through class C. However, the advantage of shifting to lower conduction angles is that efficiency increases.

Transconductance Model

Figure 3.1 shows the simplified circuit for current source mode CMOS PAs. This simplified circuit will be used for the discussion of these types of PAs. The circuit shows a blocking capacitor, C_{BLOCK}, which prevents the DC current from flowing to the output. An output filter formed by L_0 and C_0 has been added in order to remove the harmonics of the output signal.

In current source PAs the transistor can be regarded as a current source with a transconductance model showing two main limits for the current. These two limits are shown in Fig. 3.2 for a strong nonlinear transconductance model, the cutoff region for a gate bias below the threshold voltage (V_{TH}) and the open channel

H. Solar Ruiz and R. Berenguer Pérez, *Linear CMOS RF Power Amplifiers*,
DOI: 10.1007/978-1-4614-8657-2_3,

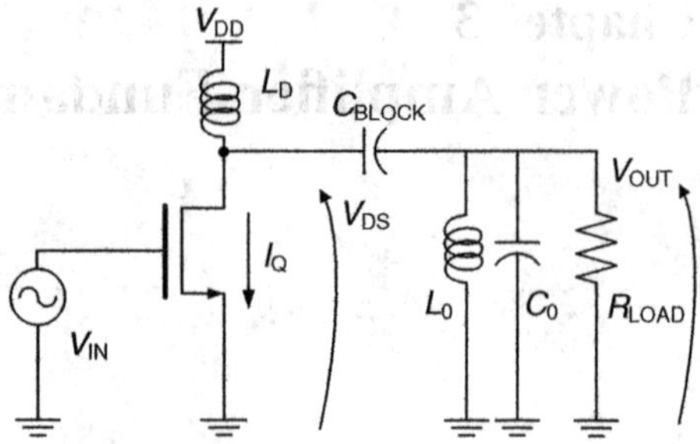

Fig. 3.1 Simplified version of a current source PA in CMOS

condition at which the output current reaches its maximum (I_{MAX}). Although this is an idealized version of actual transconductance behavior, the strong nonlinear model serves to explain how current source PAs are usually loaded and why there is an optimum load for the best PA performance.

If the strong nonlinear transconductance curve is followed, the quiescent current (I_Q) and maximum linear amplitude of the drain current are shown in Fig. 3.2. In this case, the gate bias (V_Q) is chosen so that the I_Q is in the middle point between zero and I_{MAX}. In such a situation the drain current will show linear swings until the current peak reaches zero or I_{MAX}; beyond these levels the current would start to clip.

Along with this, considering the transistor biased with V_{DD} at the drain, and in order to use the transistor to its full linear capacity, the optimum value for the load is obtained from (3.1) and it is defined as the load line match.

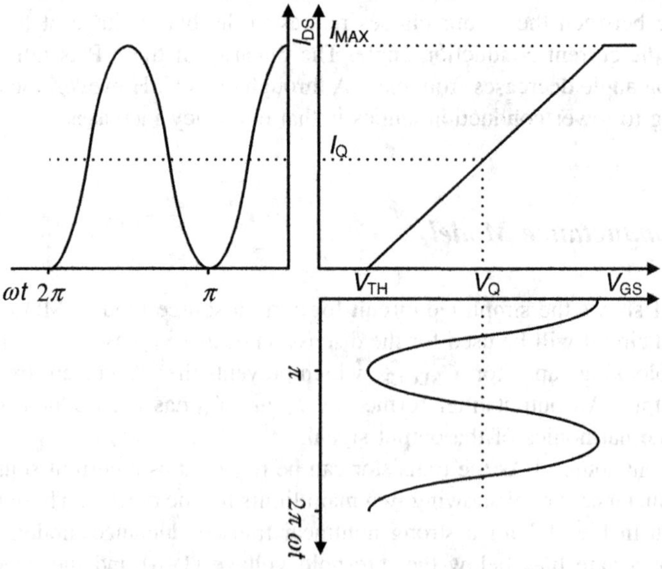

Fig. 3.2 Gate voltage and drain current swings based on the strong nonlinear model

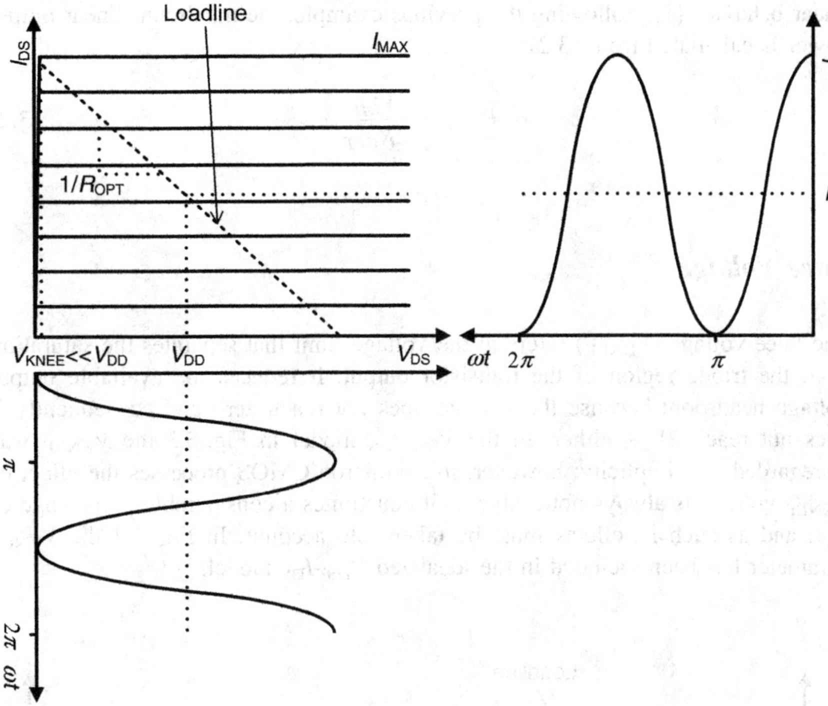

Fig. 3.3 Maximum linear drain current and voltage swings based on the load line match

$$R_{OPT} = \frac{V_{DD}}{I_{MAX}/2} \tag{3.1}$$

This value is called optimum because it is the load that provides the best linearity to the PA. This is because the drain current and voltage simultaneously swing to their maximum linear capacity, as shown in the idealized V_{DS}–I_{DS} model in Fig. 3.3. It can be observed that the drain current has reached its maximum linear swing, from 0 to I_{MAX}, and at the same time the drain voltage value, thanks to R_{OPT}, swings from V_{DD} to $2V_{DD}$, which is also its maximum linear swing.

It is important to note that the performance of the PA is measured in terms of its efficiency and the maximum linear output power levels that it can reach. Hence the maximum linear P_{OUT} is the main parameter to optimize. This is different to other type of amplifiers in which gain is the main parameter and thus the output load is matched to the conjugate of the output impedance of the amplifier. However, the conjugate match is not desirable for a PA because the transistor will not be used to its full linear capacity. In order to utilize the maximum linear drain current and voltage swings of the transistor, a lower value of load resistance must be selected. Although the load line match presents less power gain, it always gives the best linear behavior [1]. Following the previous example, the maximum linear output power is calculated from (3.2).

$$P_{OUT} = \frac{V_{DD}^2}{2R_{OPT}} \tag{3.2}$$

Knee Voltage

The knee voltage (V_{KNEE}) refers to the voltage limit that separates the saturation from the triode region of the transistor output. It reduces the available output voltage headroom because the voltage does not reach zero and consequently it does not reach $2V_{\mathrm{DD}}$ either. In the V_{DS}–I_{DS} model in Fig. 3.3 the V_{KNEE} was disregarded for simplicity; however, in submicron CMOS processes the effect of V_{KNEE} voltage is always noticeable as it constitutes a considerable percentage of V_{DD} and as such its effects must be taken into account. In Fig. 3.4 the V_{KNEE} parameter has been included in the idealized V_{DS}–I_{DS} model.

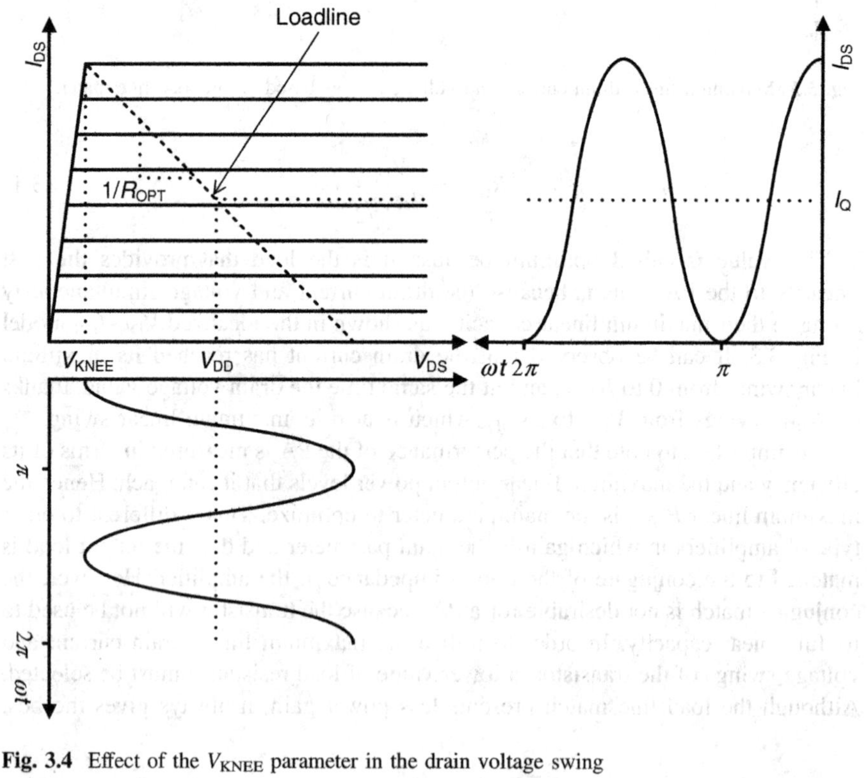

Fig. 3.4 Effect of the V_{KNEE} parameter in the drain voltage swing

If V_{KNEE} is now taken into account, R_{OPT} must be recalculated as in (3.3).

$$R_{OPT} = \frac{V_{DD} - V_{KNEE}}{I_{MAX}/2} \tag{3.3}$$

As the output voltage swing is reduced from V_{DD} to V_{DD}–V_{KNEE}, the maximum linear output power is directly affected. In this case, the maximum linear output power is (3.4). The V_{KNEE} imposes an extra limitation on the maximum drain voltage headroom; its effect will be quantified in Chap 4.

$$P_{OUT} = \frac{(V_{DD} - V_{KNEE})^2}{2R_{OPT}} \tag{3.4}$$

Class A PAs

In class A PAs, the gate is precisely the same biasing of Fig. 3.2, i.e. the gate is biased in the middle of the transconductance curve. As shown in Fig. 3.5, following the strong nonlinear transconductance model two different input drive levels are depicted: an intermediate level in a solid red line and the maximum level before clipping in a solid black line. As can be observed, PA performance is linear for all the intermediate values and the maximum linear output power that can be achieved is again given by (3.4).

In class A PAs there is a continuous current consumption that is practically independent of the output power. The class A mode is always consuming power because the transistor is in the active region for the entire input cycle. Drain efficiency in this case is poor, being 50 % of the theoretical maximum. In fact, maximum drain efficiency, by ignoring in this case the effect of V_{KNEE}, can be calculated as (3.5).

$$\eta(\%) = 100\frac{P_{OUT}}{P_{DC}} = 100\frac{V_{DD}^2/2R_{OPT}}{V_{DD}I_{DC}} = 100\frac{V_{DD}I_{DC}}{2V_{DD}I_{DC}} = 50\% \tag{3.5}$$

The power capability of the class A operation mode, where the knee voltage effect is ignored, is given by (3.6). As will be seen when compared to the other PA classes, the class A mode has a relatively high device stress compared to other different architectures.

$$P_N = \frac{P_{OUT}}{v_{DS,\max} i_{DS,\max}} = \frac{V_{DD}I_{DC}/2}{2V_{DD}2I_{DC}} = \frac{1}{8} \tag{3.6}$$

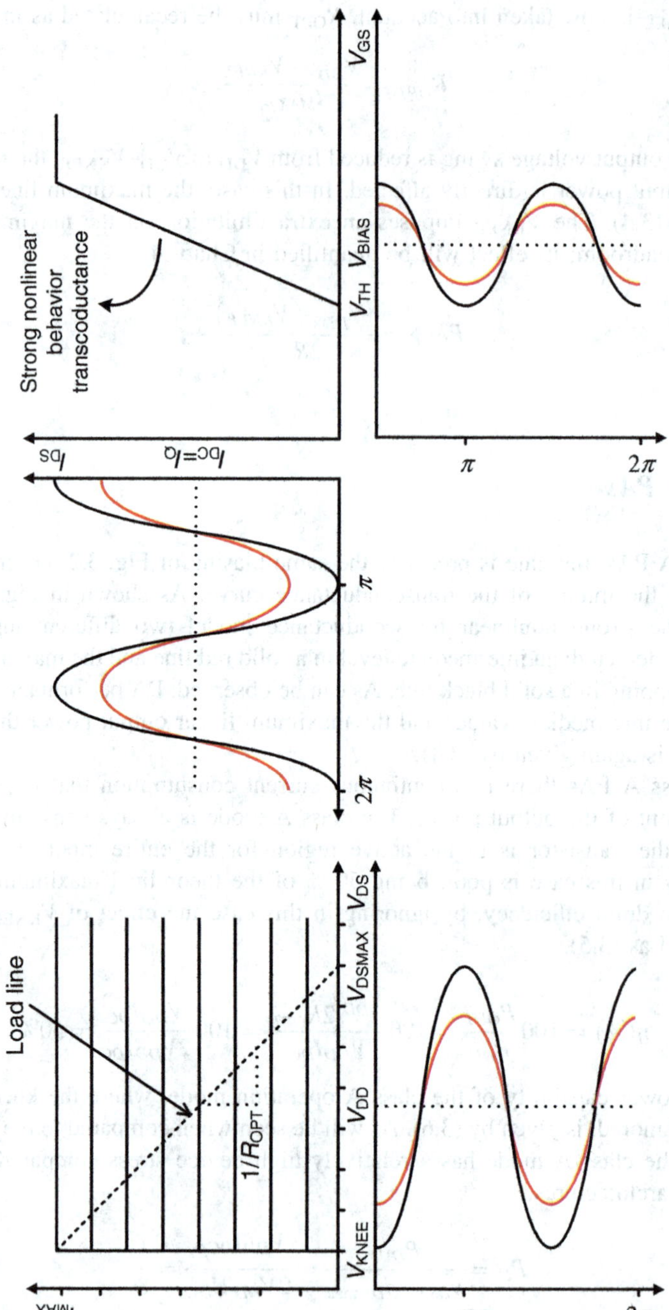

Fig. 3.5 Input and output signals for a class A PA

Class AB PAs

In class AB PAs, the biasing voltage at the gate is lowered so that the I_Q is also reduced. If the strong nonlinear model for the transconductance is assumed, the drain current waveform and its DC and fundamental components are those in Fig. 3.6. Again, two different input drive levels are depicted: an intermediate level in the solid red line and the maximum level before clipping at top in the solid black line. As can be observed, for low drive voltages the class AB PA performs like class A PAs. However, when the gate voltage is high enough the drain current clips at zero. In such situations, the current would present harmonics at the output but they are shorted to ground by the aforementioned LC filter so that only the first harmonic (I_1) flows through the load, as can be seen in the dashed line in Fig. 3.6. The amplitude of the fundamental harmonic depends on the angle $\alpha/2$, i.e. the angle at which the drain current equals zero or the conduction angle. Based on this strong nonlinear model, the fundamental harmonic can be expressed as in (3.7) [1].

$$I_1 = \frac{I_{MAX}}{2\pi}\frac{\alpha - \sin(\alpha)}{1 - \cos(\alpha/2)} \tag{3.7}$$

The value of the conduction angle depends on the different amplitudes that the drain current takes for different input drive levels and thus different values for I_1 are obtained. Specifically, the value of the conduction angle just before the drain current starts to clip at I_{MAX} is taken for comparison among the different linear classes; it appears as $\alpha_0/2$ in Fig. 3.6. The angle $\alpha_0/2$ for a class AB PA varies from π, for an operation mode close to a class A PA, to $\pi/2$ for a biasing close to a class B PA.

On the other hand, as mentioned in Chap. 2, the current consumption of a class AB PA is not constant and the current consumption variation is more noticeable as the biasing gets closer to a class B PA. As observed in Fig. 3.6, when the drain current clips at zero, I_{DC} separates from the quiescent current. This will be also analyzed and discussed in Chap. 6 with a more realistic model. For the strong nonlinear model, the different values that the I_{DC} takes are given by (3.8).

$$I_{DC} = \frac{I_{MAX}}{2\pi}\frac{2\sin(\alpha/2) - \alpha\cos(\alpha/2)}{1 - \cos(\alpha/2)} \tag{3.8}$$

In addition, the power capability of the class AB amplifiers is slightly better than that of the class A amplifier. This is reflected in Fig. 3.6 in the fact that the R_{OPT} must be decreased slightly in order to accommodate the amplitude of harmonic I_1, which is higher than the amplitude of the drain current. In fact the expression in (3.7) is always greater than 1/2 for any $\alpha_0/2$ in a class AB biasing, with a maximum of 0.54. The power capability of a class AB PA is (3.9), with a maximum of $P_N = 0.135$.

$$P_N = \frac{P_{OUT}}{v_{DS,\max} i_{DS,\max}} = \frac{V_{DD} I_1/2}{2V_{DD}I_{MAX}} = \frac{1}{4}\frac{I_1}{I_{MAX}} = \frac{1}{8\pi}\frac{\alpha_0 - \sin(\alpha_0)}{1 - \cos(\alpha_0/2)} \tag{3.9}$$

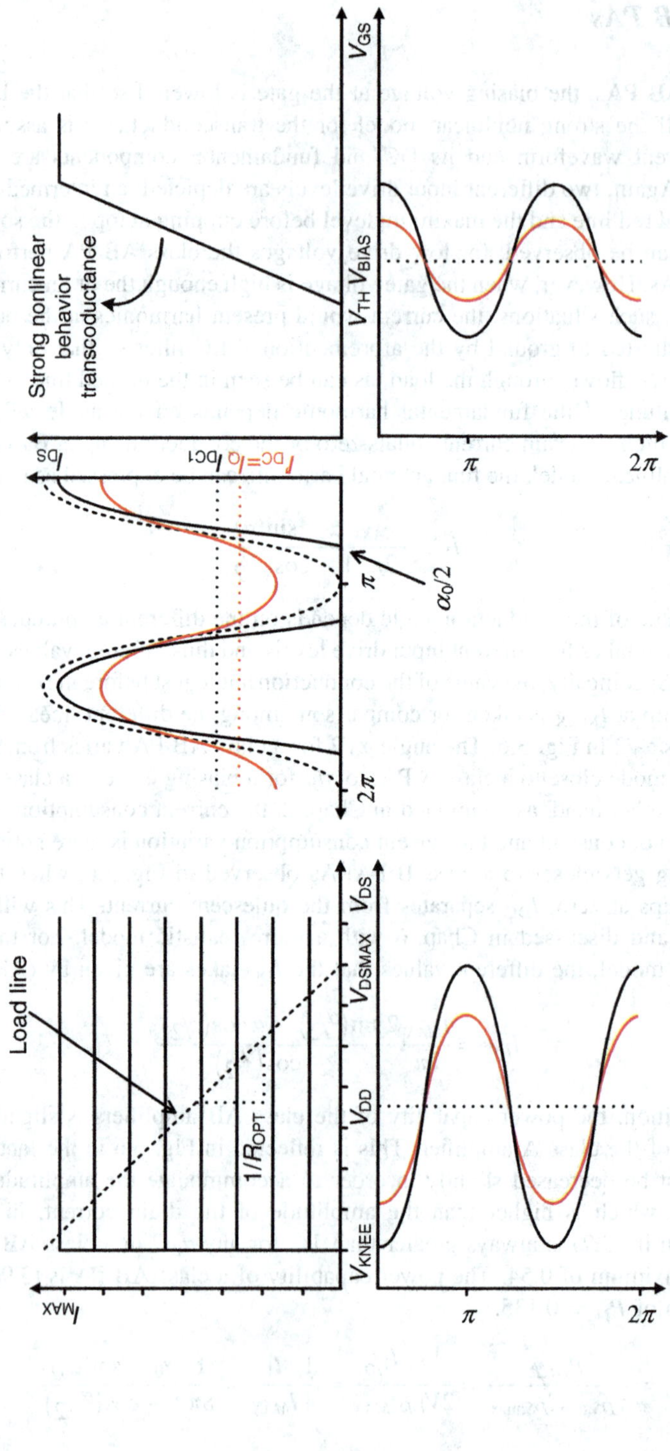

Fig. 3.6 Input and output signals for a class AB PA

The theoretical maximum drain efficiency is given by (3.10) and is between drain efficiency given for a class A PA (50 %) and drain efficiency of a class B PA (78.5 %).

$$\eta(\%) = 100\frac{P_{OUT}}{P_{DC}} = 100\frac{1}{2}\frac{\alpha_0 - \sin(\alpha_0)}{2\sin(\alpha_0/2) - \alpha_0\cos(\alpha_0/2)} \tag{3.10}$$

Finally, it must be noted that, based on the strong nonlinear transconductance of Fig. 3.2, the class A PA would show the best linearity performance among the conventional current source PAs. However, in a more realistic strong–weak nonlinear transconductance shown as a dashed line in Fig. 3.7, the threshold between the linear and nonlinear regions of the transistor is not so clear.

For the strong–weak nonlinear model a class A PA shows early compression for the drain current, as shown in Fig. 3.7 by the dashed line, and so its actual linearity is degraded. In such a case, it is possible to provide higher linearity to the PA if the biasing at the gate is reduced, i.e. from a class A to a class AB PA. The specific optimum linearity biasing may change from one CMOS process to another because the transistor may show different nonlinear characteristics of transconductance. However, this sweet spot can be found for any CMOS process by circuit simulation and therefore class AB PAs provide the best linearity among current source amplifiers, and they also offer better efficiency than the class A PAs [1].

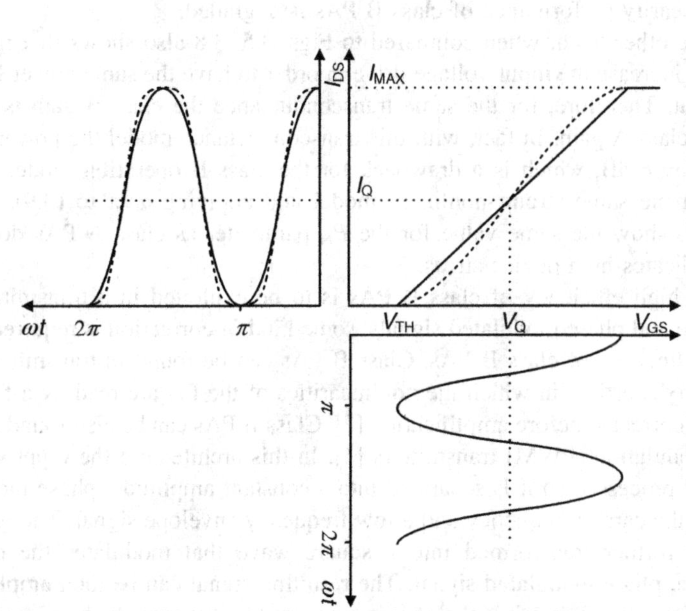

Fig. 3.7 Gate voltage and drain current swings of a class A PA based on two models of the transconductance *curve*: the strong nonlinear model (*solid line*) and the strong–weak nonlinear model (*dashed line*)

Class B PAs

In class B PAs the gate bias voltage is dropped to zero so that there is no quiescent current, as shown in Fig. 3.8. Again, the transconductance has been idealized to an ideal strong nonlinear performance, and two different input drive levels are also depicted: an intermediate level in the solid red line and the maximum level before clipping in the solid black line. The filtered fundamental output current and voltage waveforms are also shown in Fig. 3.8. As it can be observed, the current consumption is not constant but rather it is dependent on the drain current level.

Additionally, since the current consumption is reduced, there is greater efficiency than there is with class A or class AB PAs. The maximum linear output power that can be achieved is, again, given by (3.4). Drain efficiency in this case has a theoretical maximum of 78.5 % (3.11). This value is obtained by means of (3.10), where the conduction angle $\alpha_0 = \pi$ [1].

$$\eta(\%) = 100\frac{P_{OUT}}{P_{DC}} = 100\frac{V_{DD}^2/2R_{OPT}}{V_{DD}I_{DC}} = 100\frac{V_{DD}I_1}{2V_{DD}I_{DC}} = 78.5\% \qquad (3.11)$$

Surprisingly, class B PAs show a linear performance, i.e. a 3 dB input power reduction results in a 3 dB output power reduction. However, the reason is that the strong nonlinear transconductance curve has been followed in the analysis. When a more realistic model, such as the strong–weak nonlinear model, is applied the actual linearity performance of class B PAs is degraded.

On the other hand, when compared to Figs. 3.5, 3.8 also shows that it is necessary to increase the input voltage drive in order to have the same power levels at the output. Therefore, for the same transconductance the class B gain is smaller than the class A gain. In fact, with this transconductance model the power gain is reduced by 6 dB, which is a drawback for the class B operation mode. Finally, assuming the same strong nonlinear model and applying α_0/π to (3.9), class B amplifiers show the same value for the P_N parameter as class A PAs do, which again indicates high device stress.

If the high efficiency of class B PAs is to be exploited in a transmitter with amplitude and phase modulated signals, some kind of correction is required due to the poor linearity of class B PAs. Class B PAs can be found in transmitters with digital predistortion, in which the nonlinearities of the PA are read by a feedback loop and corrected before amplification [2]. Class B PAs can be also found in pulse width modulation (PWM) transmitters [3]. In this architecture the input signal is baseband processed so it is separated into a constant amplitude, phase modulated signal at the carrier frequency and a low frequency envelope signal. The envelope signal is further transformed into a square wave that modulates the constant amplitude, phase modulated signal. The resulting signal can be then amplified by the nonlinear PA. The original signal is recovered by filtering the harmonics at the output of the PA. In [3] several class B PAs are combined to reach watt level output power.

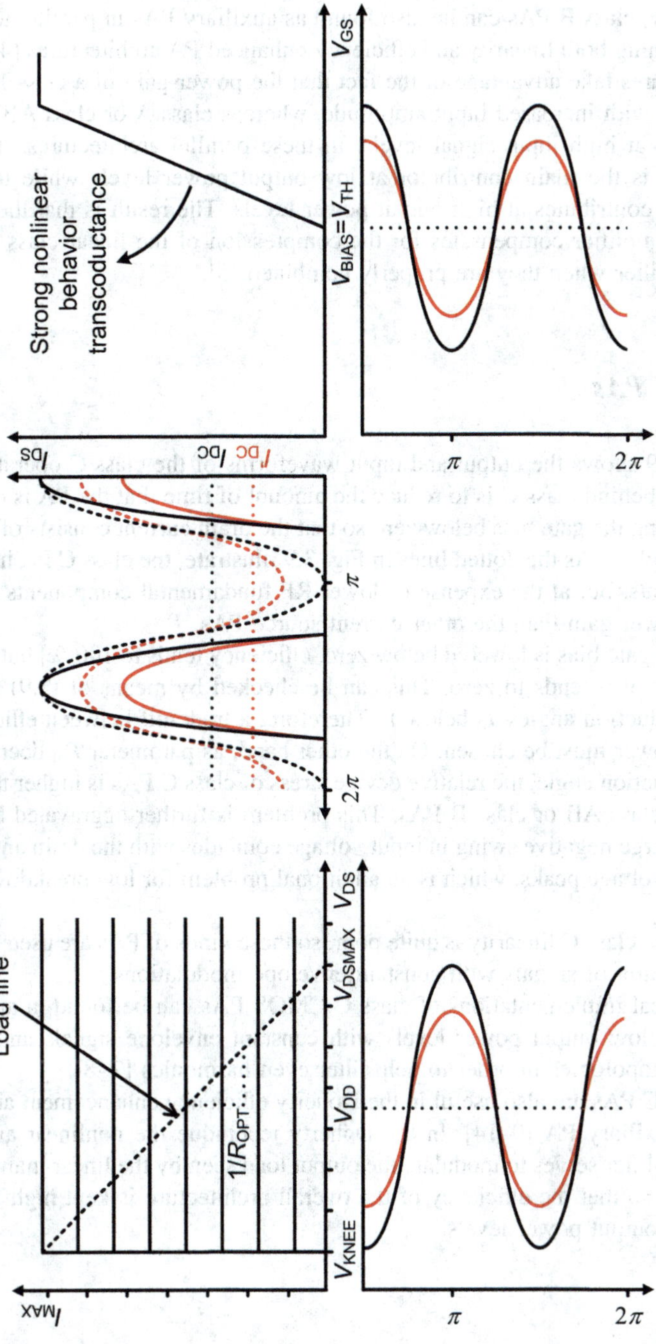

Fig. 3.8 Input and output signals for a class B PA

Finally, class B PAs can be also found as auxiliary PAs in parallel to class AB PAs, forming both linearity and efficiency-enhanced PA architectures [4–6]. These architectures take advantage of the fact that the power gain of a class B amplifier increases with increased input amplitude, whereas class A or class AB amplifiers compress at high input signal levels. In these parallel architectures, the class A amplifier is the main contributor at low output power levels while the class B amplifier contributes at high output power levels. The result is that the nonlinear class B amplifier compensates for the compression of the linear class A or class AB amplifier when they are properly combined [5].

Class C PAs

Figure 3.9 shows the output and input waveforms of the class C operation mode. The idea behind class C is to reduce the amount of time that the PA is conducting by reducing the gate bias below zero so that the drain current consists of a periodic train of pulses. As the dotted lines in Fig. 3.9 illustrate, the class C PA has low DC components, but at the expense of lower RF fundamental components and hence lower power gain than the other current source PAs.

As the gate bias is lowered below zero, efficiency tends to 100 %, but the power capability also tends to zero. This can be checked by means of (3.9) and (3.10) with conduction angles α_0 below π. Therefore, a trade-off between efficiency and output power must be chosen. On the other hand, as parameter P_N decreases with the conduction angle, the relative device stress of class C PAs is higher than that of class A, class AB or class B PAs. This problem is further aggravated by the fact that the large negative swing in input voltage coincides with the drain and collector positive voltage peaks, which is an additional problem for low breakdown voltage devices.

Finally, class C linearity is quite poor, so these kinds of PAs are used mainly for amplification of signals with constant envelope modulations.

Practical implementations of class C CMOS PAs can be found in applications requiring low output power levels with constant envelope signals and with differential topologies in order to help filter even harmonics [7, 8].

Class C PAs are also useful in the Doherty efficiency enhancement architecture as the auxiliary PA [9–14]. In the Doherty technique the nonlinear auxiliary or peak amplifier serves to modulate the output load seen by the linear main or carrier amplifier so that the efficiency of the overall architecture is kept high for a wide range of output power levels.

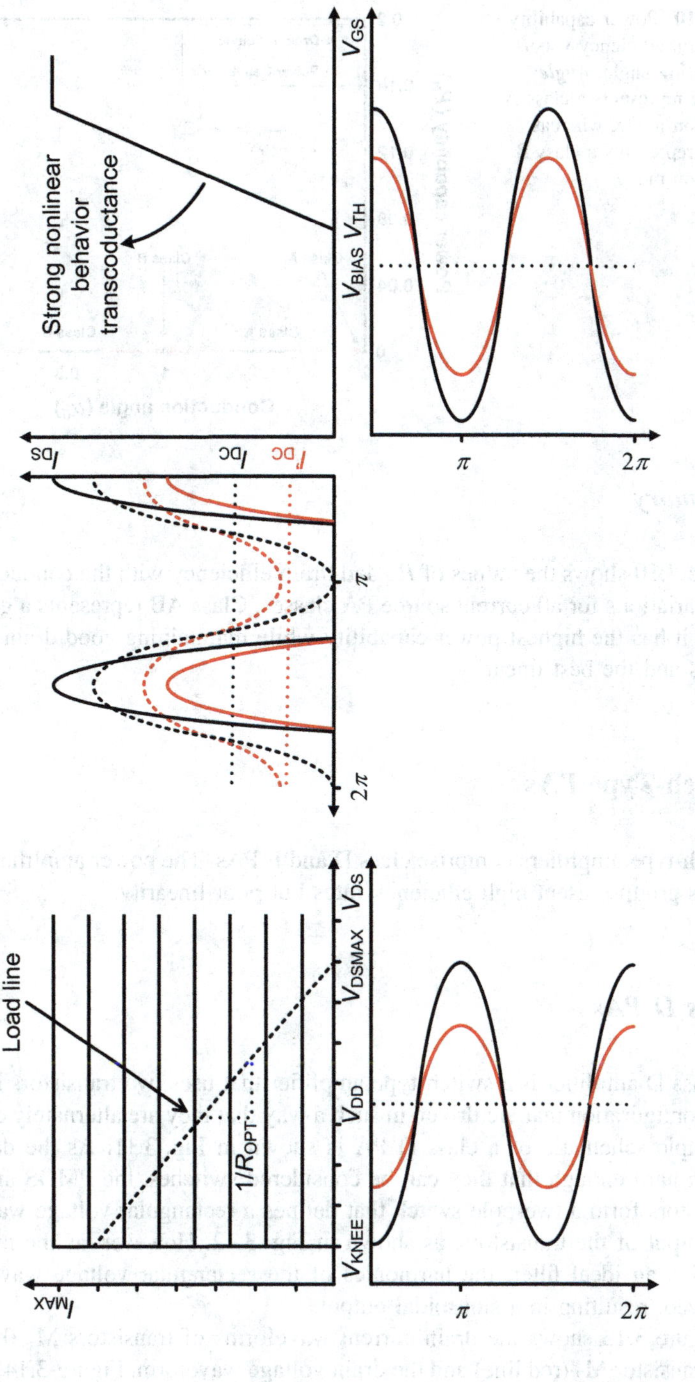

Fig. 3.9 Input and output signals for a class C PA

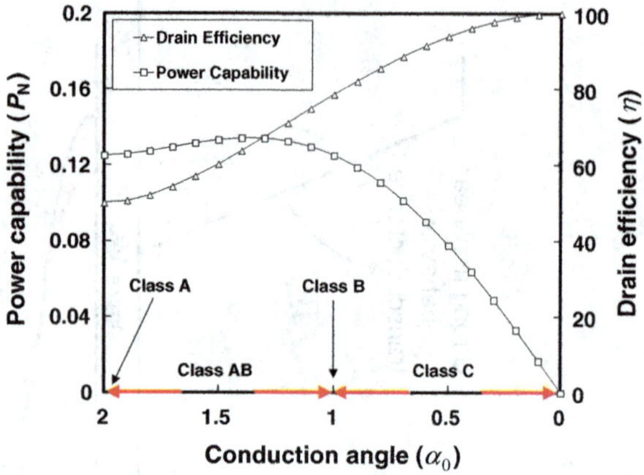

Fig. 3.10 Power capability and drain efficiency versus conduction angle. *Angle* $\alpha = 2\pi$ represents a class A operation mode, whereas $\alpha = \pi$ represents a class B operation mode

Summary

Figure 3.10 shows the values of P_N and drain efficiency with the conduction angle (α_0) variations for all current source PA classes. Class AB represents a good trade-off as it has the highest power capability while maintaining good drain efficiency values and the best linearity.

Switch-Type PAs

Switch-type amplifiers comprise class D and E PAs. The power amplifiers that fall in this group present high efficiency rates but poor linearity.

Class D PAs

A Class D amplifier is a switch-type amplifier that uses two transistors in a push–pull configuration that are driven in such a way that they are alternately on and off. A simple schematic of a class D PA is shown in Fig. 3.11. As the devices are driven hard enough that they can be considered switches, the PMOS and NMOS transistors form a two-pole switch that defines a rectangular voltage waveform at the output of the transistors, as shown in Fig. 3.12. However, as the load circuit contains an ideal filter, the harmonics of the rectangular voltage waveform are removed, resulting in a sinusoidal output.

Figure 3.13 shows the drain current waveforms of transistors M_1 (black line) and transistor M_2 (red line) and the drain voltage waveform. Figure 3.14 shows the voltage and current waveforms flowing into the load. It must be noted that the

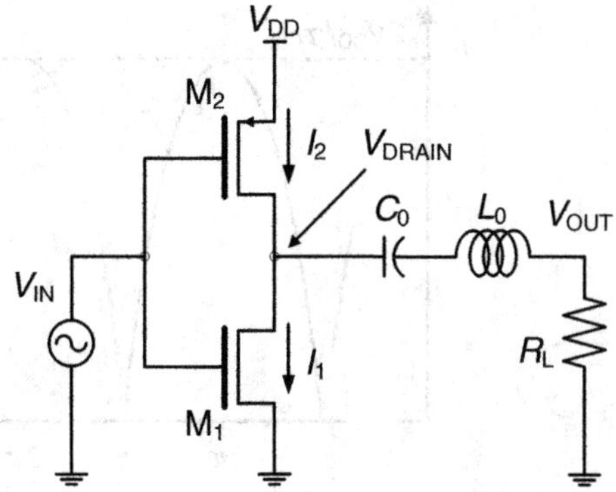

Fig. 3.11 Basic circuit of a generic class D amplifier

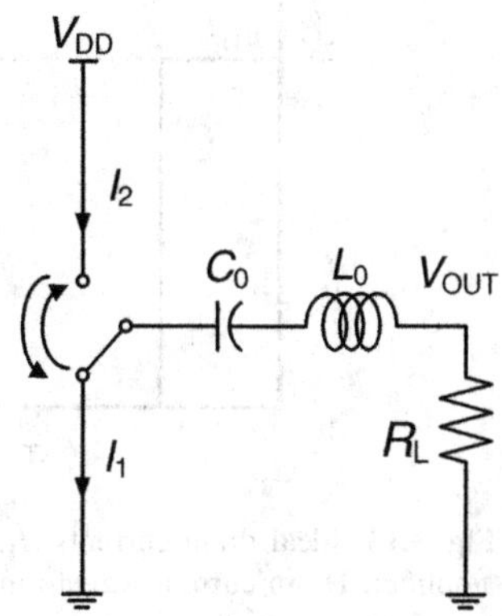

Fig. 3.12 Equivalent ideal circuit of the generic class D amplifier

amplitude of the fundamental component of a square wave voltage is $4/\pi$ times the amplitude of the square wave.

The main virtue of class D is its high power capability, i.e. relatively low device stress. The power capability can be extracted from Figs. 3.13 and 3.14 for each transistor of this architecture (3.12) [15]. Therefore, the class D operation mode shows higher power capability relative to class A through C.

$$P_N = \frac{P_{OUT}/2}{v_{DS,\max} i_{DS,\max}} = \frac{1}{2}\frac{\frac{\left(\frac{2V_{DD}}{\pi R_L}\right)^2}{2} R_L}{V_{DD}\frac{2V_{DD}}{\pi R_L}} = \frac{1}{2\pi} \approx 0.16 \qquad (3.12)$$

In class D PAs, if transistors operated as ideal switches, efficiency would be 100 % as can be observed in (3.13), in which I_{DC} is the average of the I_1 current waveform, as given in (3.14).

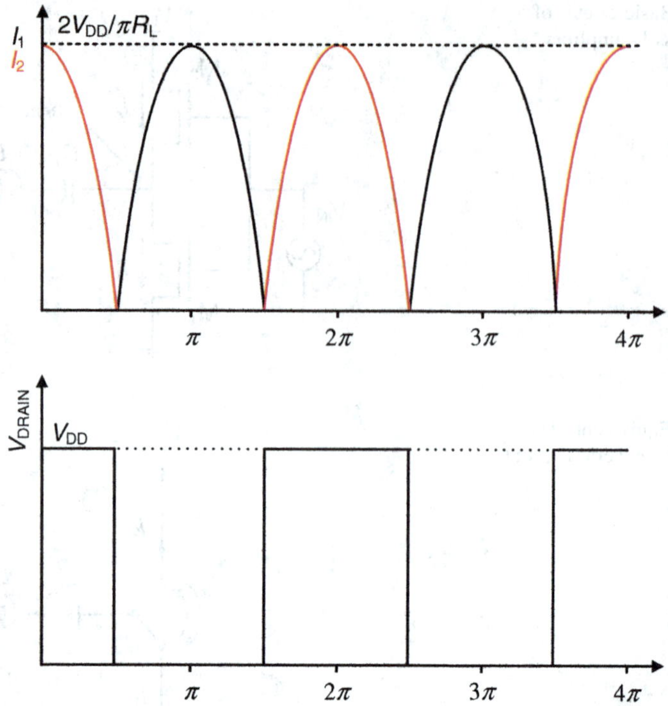

Fig. 3.13 Ideal drain currents (I_1, I_2) and voltage (V_{DRAIN}) waveforms of a push–pull class D amplifier. Drain current waveforms are depicted in *black* and *red* for each transistor

$$P_N = 100\frac{P_{OUT}}{P_{DC}} = 100\frac{\frac{\left(\frac{2V_{DD}}{\pi R_L}\right)^2}{2} R_L}{V_{DD}\frac{2V_{DD}}{\pi^2 R_L}} = 100\% \tag{3.13}$$

$$I_{DC} = avg(I_1) = \frac{2V_{DD}}{\pi^2 R_L} \tag{3.14}$$

However, limited switch performance and the large parasitic drain-source capacitance of transistors cause efficiency to drop well below 100 %. In fact, switch operation modes in CMOS processes tend to work well only at frequencies substantially below their transistor f_T. In addition, class D PAs alone cannot be used for applications in which good linearity is required, as their performance is strongly nonlinear.

Practical implementations of class D CMOS PAs can be found in the handling of high output power levels at low frequencies in audio applications [16–20]. However, as the switching performance of deep submicron CMOS transistors improves, class D PAs have attracted the attention of linear PA designers at the GHz range. The transmitter architectures in which class D PAs are implemented

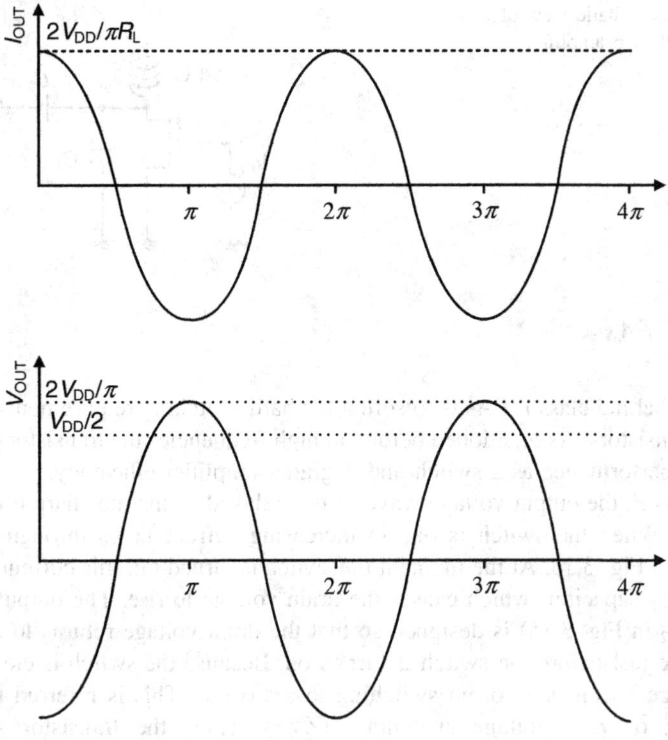

Fig. 3.14 Filtered output current (I_{OUT}) and voltage (V_{OUT}) waveforms at the load

and show high performance in the GHz range are the outphasing [21–27] and polar architectures [28–31]. In these architectures, performance enhancement is pursued by using nonlinear PAs for amplitude and phase modulated signals that would require high linearity. In order to achieve this, the input signal is first baseband processed and decomposed into several signals so that these resulting signals can be applied to nonlinear PAs.

In the case of the outphasing architecture, the input signal is decomposed into two amplitude-constant, phase-modulated signals at the carrier frequency. Once the baseband processed signals have been amplified, they are finally recombined at the output of the PA in order to restore the original amplitude and phase modulated signal. The main drawback of this architecture is the need for an efficient power combiner at the output, which is difficult to integrate in CMOS [32].

The polar architecture separates the input signal into an amplitude-constant, phase-modulated signal at the carrier frequency and the low frequency envelope signal. The envelope signal serves to modulate the supply voltage of nonlinear PAs, and the issues related to PA supply modulation are overcome by means of different solutions, such as digital modulation [28, 30], PWM [29], ΣΔ modulation [31] or pulse density modulation (PDM) [33].

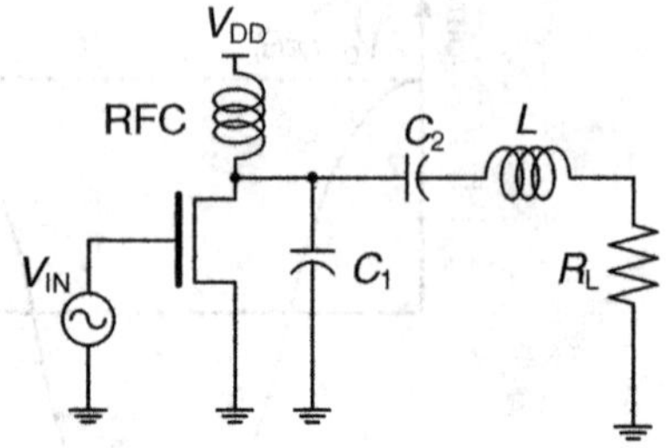

Fig. 3.15 Schematic view of a generic class E amplifier

Class E PAs

The idea behind class E PAs is to soften the hard switching requirements imposed on the transistors. As mentioned before, at high frequencies the transistor parasitics spoil its performance as a switch and degrades amplifier efficiency.

In class E, the output voltage waveform is relaxed so that no sharp transition is required. When the switch is on, an increasing current flows through the RFC inductor in Fig. 3.15. At the moment the switch is turned off, this current is driven into the C_1 capacitor, which causes the drain voltage to rise. The output network (L and C_2 in Fig. 3.15) is designed so that the drain voltage returns to zero with zero slope just before the switch is turned on. Because the switch is closed when the voltage becomes zero, no switching losses occur. This is referred to as soft switching or zero voltage switching, (ZVS). Thus, the transistor switch-on requirements of the class E PA are relaxed compared to the class D requirements. In fact, the parasitic output capacitance of the transistor can be part of the output network, which is a significant advantage, especially in CMOS [32]. This output network also serves to filter the fundamental component to the output

Values for the aforementioned components are given by the following expressions (3.15–3.17), where Q is the quality factor of the output network [34].

$$L = \frac{QR_L}{\omega} \tag{3.15}$$

$$C_1 \approx \frac{1}{\omega(5.447R_L)} \tag{3.16}$$

$$C_2 \approx C_1\left(\frac{5.447}{Q}\right)\left(1+\frac{1.42}{Q-2.08}\right) \tag{3.17}$$

Figure 3.16 shows example values of the voltage and current waveforms. The voltage and current waveforms are depicted in black for the transistor whereas the current flowing through C_1 appears in red.

In the class E operation mode, the theoretical maximum efficiency is 100 %. However, the switch-off requirements of the drain current (the switch turns off when current is nearly at its maximum) can still spoil efficiency if the switch is not

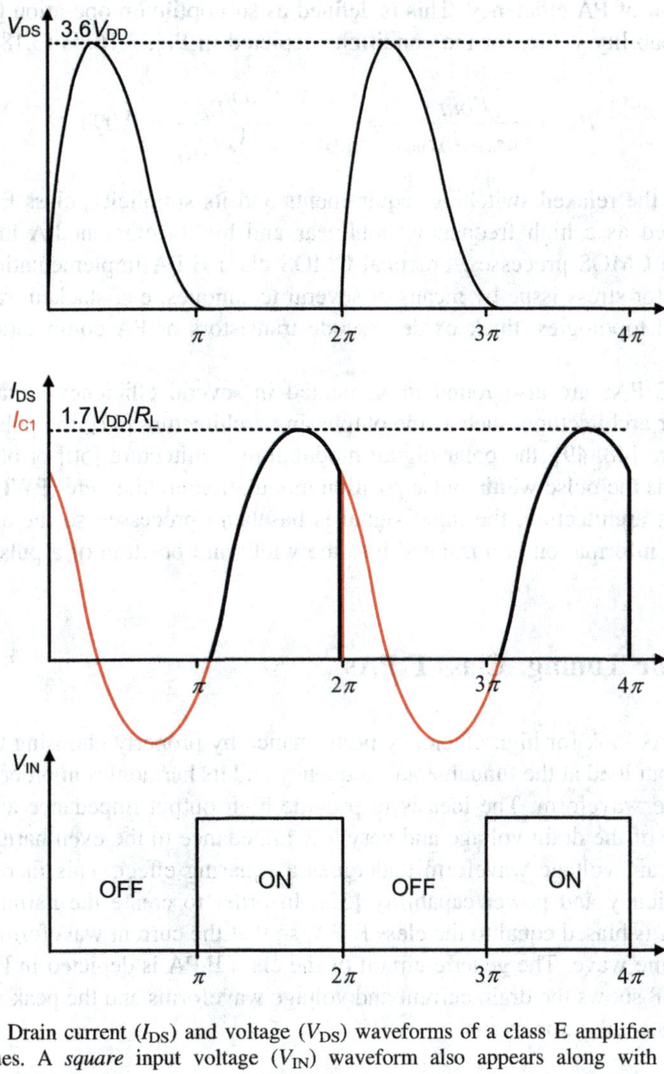

Fig. 3.16 Drain current (I_{DS}) and voltage (V_{DS}) waveforms of a class E amplifier with typical peak values. A *square* input voltage (V_{IN}) waveform also appears along with the current waveform of the shunt capacitor (I_{C1}), in solid *red*

fast enough [35], so that even in class E a fast switch is important in order to keep the efficiency high.

Moreover, class E PAs have low power capability. The reason for that can be observed in Fig. 3.16, in the fact that the peak drain voltage is large. Therefore, class E stresses the device more than any other previously described PA class; this is a primary drawback of class E PAs, especially in CMOS breakdown voltage limited processes. The high peak drain voltage can be relaxed, but it comes with a degradation of PA efficiency. This is defined as sub-optimum operation [36]. The power capability value, for the conditions depicted in Fig. 3.16, is (3.18).

$$P_N = \frac{P_{OUT}}{v_{DS,\max} i_{DS,\max}} = \frac{0.577 V_{DD}/R_L}{3.6 V_{DD} 1.7 V_{DD}/R_L} = 0.094 \quad (3.18)$$

Due to the relaxed switching requirements and its simplicity, class E PAs are widely used as a high frequency, nonlinear and highly efficient PA in modern submicron CMOS processes. Practical CMOS class E PA implementations relax the transistor stress issue by means of several techniques, e.g. stacked transistors, differential topologies, thick oxide cascode transistors or PA combinations [37–46].

Class E PAs are also found implemented in several efficiency enhancement transmitter architectures such as the outphasing architecture [47], the polar analog architecture [48, 49], the polar digital modulation architecture [50] or other variants such as the pulse width, pulse position modulation architecture (PWPM) [51]. In this last architecture, the input signal is baseband processed so the amplitude and phase information is translated into the width and position of a pulse.

Harmonic Tuning: Class F PAs

Class F PAs look for high efficiency performance by properly choosing the value of the output load at the fundamental frequency and its harmonics in order to shape the voltage waveform. The idea is to provide high output impedance at the odd harmonics of the drain voltage and very low impedance to the even harmonics so that the drain voltage waveform undergoes a squaring effect. This increases the PA's efficiency and power capability [52]. In order to create the harmonics the class F PA is biased equal to the class B PA, so that the current waveform is a half rectified sine wave. The generic circuit of the class F PA is depicted in Fig. 3.17. Figure 3.18 shows the drain current and voltage waveforms and the peak values of the fundamental components.

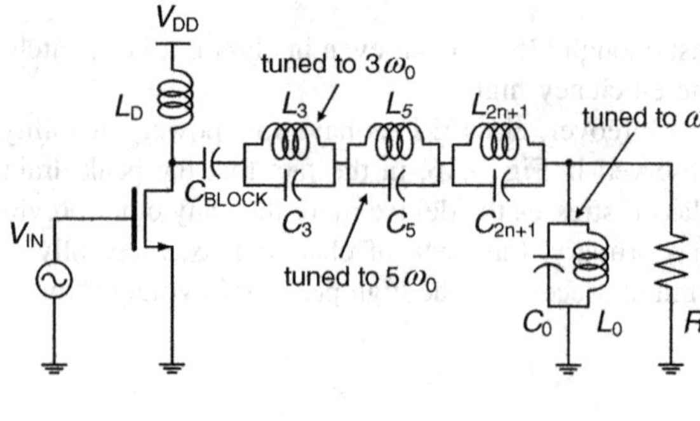

Fig. 3.17 Schematic view of a generic class F amplifier

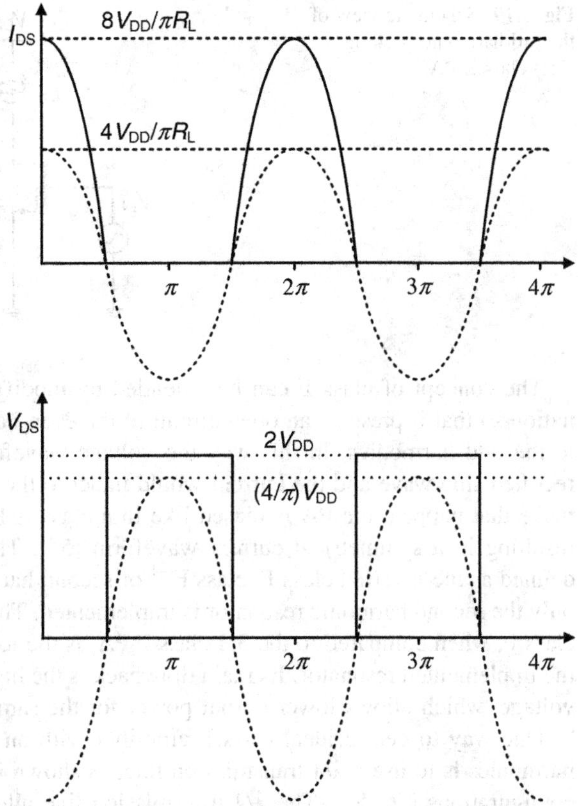

Fig. 3.18 Ideal drain current (I_{DS}) and voltage (V_{DS}) waveforms for a class F amplifier. Fundamental component of the current and voltage waveforms are also plotted in as a *dotted line*

In the circuit of Fig. 3.17, a class F power amplifier implemented with an infinite number of resonators is assumed so that it allows a perfect squaring effect of the output voltage by trapping all the odd harmonics. Class F amplifiers are sometimes designated as a switching amplifier, but this PA can only be considered a switch-mode PA when an infinite number of harmonics are added. Practical implementations of integrated class F PAs only implement the third harmonic resonator [53, 54], and thus the PA transistor has the same drive requirements as a class B amplifier [32]. This is sufficient for producing significant improvement in drain efficiency [52], and there are two main reasons for limiting the number of resonators. The first reason is the low quality factor of the integrated inductors that limit the number of resonators that can be implemented without excessively degrading the PA efficiency. The second reason is that the drain-source parasitic capacitance, which is usually large compared to the size of the transistors, causes a low-impedance path for the higher harmonics. This results in the reduced amplitude of the higher harmonics and a less pronounced square voltage. Therefore, the loading network best suited for the integration of class F PAs is based on one resonator only, and this is usually referred to as third harmonic peaking (3H), whose circuit is shown in Fig. 3.19.

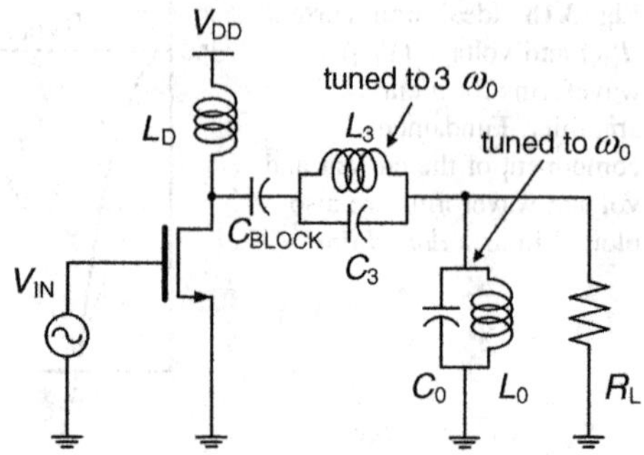

Fig. 3.19 Schematic view of the 3rd harmonic peaking (3H) class F PA

The concept of class F can be extended by modifying the output load termination so that it presents an open circuit to the even harmonics and a short circuit to the odd harmonics. In this way the voltage waveform would turn into a half rectified sine wave and the current would undergo the squaring effect. In order to make that happen, the PA is biased like in a class A PA in overdrive conditions, resulting in a symmetrical current waveform [55]. This class F configuration is defined as the inverted class F, class F^{-1} or second harmonic peaking (2H), where only the second harmonic resonator is implemented. The main advantage of the 2H class F, when compared to the 3H class F PA, is the lower operating frequency of the implemented resonator. Its main drawback is the higher peak value of the drain voltage, which allows lower output power for the same device stress.

One way to get an ideal class F amplifier with an infinite number of trapped harmonics is to use a $\lambda/4$ transmission line, as shown in Fig. 3.20 in two possible configurations [56, 57]. The $\lambda/4$ transmission line allows low impedance for the even harmonics and high impedance for the odd ones, such that the output voltage is shaped into a square wave. However, the use of a $\lambda/4$ transmission line will increase the area required to implement the amplifier, so it can only be reasonably integrated at high frequencies [32]. The issue of the length of the $\lambda/4$ transmission line can be reduced if it is approximated with lumped elements, but this comes with the problem of reduced efficiency due to the low quality factors of the integrated standard inductors [58, 59]. In [59] the outphasing efficiency enhancement architecture has been also implemented.

The theoretical efficiency of 3H class F PAs is 88.4 % [52], and it would be 100 % for an ideal class F PA with perfect square drain voltage and half rectified current waveforms. However, actual efficiency is reduced because of the impossibility of perfectly squaring the voltage waveform, which is due to the parasitic capacitance and the losses introduced by the networks that are required to shape it.

However, assuming that the voltage waveforms are perfectly squared, the power capability of this class PA, which is given by (3.19), is equal to class D PA.

$$P_N = \frac{P_{OUT}}{v_{DS,\max} i_{DS,\max}} = \frac{\frac{4/\pi\, V_{DD}\; 4/\pi\, V_{DD}/Z_L}{2}}{2V_{DD}\, 8/\pi\, V_{DD}/Z_L} = \frac{1}{2\pi} \approx 0.16 \tag{3.19}$$

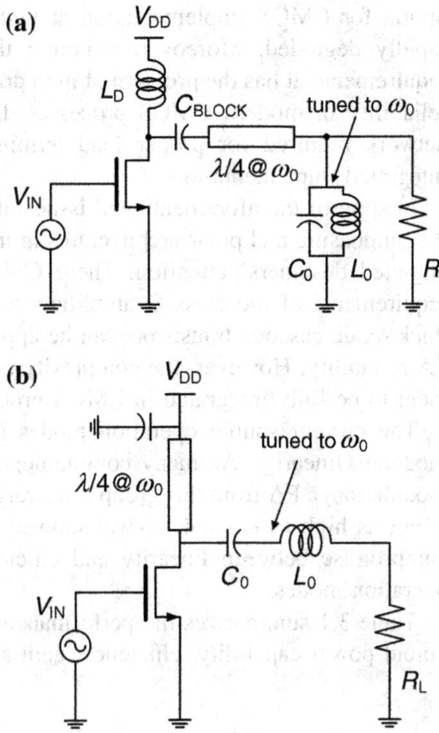

Fig. 3.20 Schematic view of a class F PA with a $\lambda/4$ transmission line

Table 3.1 Summary of the PA classes

Class	Modes	Cond. angle (%)	Output power	Max. drain efficiency (%)	Gain	Linearity
A		100	Moderate	50	Large	Good
AB	Current source	100–50	Moderate	50–78.5	Large	Good
B		50	Moderate	78.5	Moderate	Moderate
C		<50	Small	100	Small	Poor
D	Switch	50	Large	100	Small	Poor
E		50	Large	100	Small	Poor
F	SW/CS	50	Large	100	Small	Poor

Conclusions

The most interesting classes in terms of their output power capability and efficiency are the switch-mode (classes D and E) PAs and class F PAs. However, these PA classes are highly nonlinear so they cannot be directly used for amplifying spectrally efficient signals of modern communication standards. Along with this, class D PA switching requirements usually make the PA an unattractive option for CMOS implementation at high RF frequencies because efficiency is rapidly degraded. Moreover, although the class E PA relaxes the switching requirements, it has the problem of high drain peak voltages that are critical for the reliability in modern CMOS processes. In the case of class F PAs, the output network required for proper load termination degrades efficiency in practical integrated implementations.

In spite of the aforementioned issues, the combination of nonlinear PAs using the outphasing and polar architectures in modern submicron CMOS processes has attracted designers' attention. These CMOS processes overcome the switching requirements of the class D amplifiers while techniques such as the stacked or thick oxide cascode transistors can be applied to class E PAs in order to improve PA reliability. However, the complexity of these techniques makes it difficult for them to be fully integrated in CMOS processes.

The current source operation modes (classes A, AB and B) show good to moderate linearity. As many communication standards use phase and amplitude modulation, a PA from this group is a straightforward choice, even when efficiency is not as high as desired. As we showed previously, class AB presents the best compromise between linearity and efficiency in practice within current source operation modes.

Table 3.1 summarizes the performance of the different PA classes in terms of output power capability, efficiency, gain and linearity.

References

1. Cripps SC (1999) RF Power Amplifiers for Wireless Communications. Artech House, Norwood
2. Kwon DH, Hao L, Yuchun C, Tseng R, Yun C (2010) Digitally Equalized CMOS Transmitter Front-End With Integrated Power Amplifier. IEEE J Solid-State Circ 45(8):1602–1614
3. Nuyts PAJ, Francois B, Dehaene W, Reynaert P (2012) A CMOS Burst-Mode Transmitter With Watt-Level RF PA and Flexible Fully Digital Front-End. IEEE Trans Circ Syst II: Exp Briefs 59(10):613–617
4. Qing L, Sun J, YongJu S, Horie K, Itoh N, Yoshimasu T (2011) A high efficiency and high linearity power amplifier utilizing post-linearization technique for 5.8 GHz DSRC applications. In: IEEE topical conf power amplifiers for wireless and radio applications (PAWR 2011), pp 45–48
5. Yongwang D, Harjani R (2005) A high-efficiency CMOS +22-dBm linear power amplifier. IEEE J Solid-State Circ 40(9):1895–1900
6. Jian F, Shilei H, Yumei H, Zhiliang H (2010) A 2.4G-Hz CMOS power amplifier. In: IEEE international conference solid-state and integrated circuit technology (ICSICT 2010), pp 659–661
7. Natarajan K, Walling JS, Allstot DJ (2011) A class-C power amplifier/antenna interface for wireless sensor applications. In: IEEE radio frequency integrated circuits symposium (RFIC 2011), pp 1–4
8. Gupta R, Ballweber BM, Allstot DJ (2001) Design and optimization of CMOS RF power amplifiers. IEEE J Solid-State Circ 36(2):166–175

9. Elmala M, Paramesh J, Soumyanath K (2006) A 90-nm CMOS Doherty power amplifier with minimum AM-PM distortion. IEEE J Solid-State Circ 41(6):1323–1332
10. Kang J, Daekyu Y, Kyoungjoon M, Kim B (2006) A ultra-high PAE Doherty amplifier based on 0.13-mm CMOS process. IEEE Microw Wireless Compon Lett 16(9):505–507
11. Li-Yuan Y, Hsin-Shu C, Yi-Jan C (2008) A 2.4 GHz fully integrated cascode-cascade CMOS Doherty power amplifier. IEEE Microw Wireless Compon Lett 18(3):197–199
12. Wongkomet N, Tee L, Gray PR (2006) A +31.5 dBm CMOS RF Doherty power amplifier for wireless communications. IEEE J Solid-State Circ 41(12):2852–2859
13. Kaymaksut E, Reynaert P (2012) Transformer-based uneven Doherty power amplifier in 90 nm CMOS for WLAN applications. IEEE J Solid-State Circ 47(7):1659–1671
14. Chen YJE, Chih-Yun L, Tang-Nian L, Heo D (2006) A high-efficient CMOS RF power amplifier with automatic adaptive bias control. IEEE Microw Wireless Compon Lett 16(11):615–617
15. Albulet M (2001) RF power amplifiers. Noble Publishing, Atlanta
16. Guanziroli F, Bassoli R, Crippa C, Devecchi D, Nicollini G (2012) A 1 W 104 dB SNR filter-less fully-digital open-loop Class D audio amplifier with EMI reduction. IEEE J Solid-State Circ 47(3):686–698
17. Dooper L, Berkhout M (2012) A 3.4 W digital-in Class-D audio amplifier in 0.14 mm CMOS. IEEE J Solid-State Circ 47(7):1524–1534
18. Torres J, Colli-Menchi A, Rojas-Gonzalez MA, Sanchez-Sinencio E (2011) A low-power high-PSRR clock-free current-controlled Class-D audio amplifier. IEEE J Solid-State Circ 46(7):1553–1561
19. Jingxue L, Gharpurey R (2011) Design and analysis of a self-oscillating Class D audio amplifier employing a hysteretic comparator. IEEE J Solid-State Circ 46(10):2336–2349
20. Teplechuk MA, Gribben T, Amadi C (2011) True filterless Class-D audio amplifier. IEEE J Solid-State Circ 46(12):2784–2793
21. Tsai-Pi H, Choi DK, Larson LE, Asbeck PM (2007) CMOS outphasing Class-D amplifier with chireix combiner. IEEE Microw Wireless Compon Lett 17(8):619–621
22. Wei T, Hongtao X, Ravi A, Lakdawala H, Bochobza-Degani O, Carley LR, Palaskas Y (2012) A transformer-combined 31.5 dBm outphasing power amplifier in 45 nm LP CMOS with dynamic power control for back-off power efficiency enhancement. IEEE J Solid-State Circ 47(7):1646–1658
23. Landin PN, Fritzin J, Van Moer W, Isaksson M, Alvandpour A (2012) Modeling and digital predistortion of Class-D outphasing rf power amplifiers. IEEE Trans Microw Theory Techn 60(6):1907–1915
24. Hongtao X, Palaskas Y, Ravi A, Sajadieh M, El-Tanani MA, Soumyanath K (2011) A flip-chip-packaged 25.3 dBm Class-D outphasing power amplifier in 32 nm CMOS for WLAN application. IEEE J Solid-State Circ 46(7):1596–1605
25. Sungho L, Sangwook N (2010) A CMOS Outphasing Power Amplifier With Integrated Single-Ended Chireix Combiner. IEEE Trans Circ Syst II: Exp Briefs 57(6):411–415
26. Fritzin J, Jung Y, Landin PN, Handel P, Enqvist M, Alvandpour A (2011) Phase predistortion of a Class-D outphasing RF amplifier in 90 nm CMOS. IEEE Trans Circ Syst II: Exp Briefs 58(10):642–646
27. Ravi A, Madoglio P, Hongtao X, Chandrashekar K, Verhelst M, Pellerano S, Cuellar L, Aguirre-Hernandez M, Sajadieh M, Zarate-Roldan JE, Bochobza-Degani O, Lakdawala H, Palaskas Y (2012) A 2.4-GHz 20–40-MHz channel WLAN digital outphasing transmitter utilizing a delay-based wideband phase modulator in 32-nm CMOS. IEEE J Solid-State Circ 47(12):3184–3196
28. Chowdhury D, Lu Y, Alon E, Niknejad AM (2011) An efficient mixed-signal 2.4-GHz polar power amplifier in 65-nm CMOS technology. IEEE J Solid-State Circ 46(8):1796–1809
29. Nakatani T, Rode J, Kimball DF, Larson LE, Asbeck PM (2012) Digitally-controlled polar transmitter using a watt-class current-mode Class-D CMOS power amplifier and guanella reverse balun for handset applications. IEEE J Solid-State Circ 47(5):1104–1112

30. Sang-Min Y, Walling JS, Eum-Chan W, Jann B, Allstot DJ (2011) A switched-capacitor RF power amplifier. IEEE J Solid-State Circ 46(12):2977–2987
31. Jian C, Liang R, Jonsson F, Geng Y, Li-Rong Z (2012) The design of all-digital polar transmitter based on ADPLL and phase synchronized ΔΣ modulator. IEEE J Solid-State Circ 47(5):1154–1164
32. Reynaert P, Steyaert M (2006) RF power amplifiers for mobile communications. Springer, The Netherlands
33. Stauth JT, Sanders SR (2008) A 2.4 GHz, 20dBm Class-D PA with single-bit digital polar modulation in 90 nm CMOS. In: IEEE custom integrated circuits conference (CICC 2008) pp 737–740
34. Sokal NO, Sokal AD (1975) Class E-A new class of high-efficiency tuned single-ended switching power amplifiers. IEEE J Solid-State Circ 10(3):168–176
35. Lee TH (2004) The design of CMOS radio-frequency integrated circuits. Cambridge University Press, United Kingdom
36. Acar M, Annema AJ, Nauta B (2007) Variable-voltage Class-E power amplifiers. In: IEEE MTT-S international microwave symposium digital (IMS 2007), pp 1095–1098
37. Zhisheng L, Torfs G, Bauwelinck J, Xin Y, Vandewege J, Van Praet C, Spiessens P, Tubbax H, Stubbe F (2012) A 2.45-GHz + 20-dBm fast switching Class-E power amplifier with 43% PAE and a 18-dB-wide power range in 0.18-μm CMOS. IEEE Trans Circ Syst II: Exp. Briefs 59(4):224–228
38. Lee O, An KH, Kim H, Lee DH, Han J, Yang KS, Lee CH, Kim H, Laskar J (2010) Analysis and design of fully integrated high-power parallel-circuit Class-E CMOS power amplifiers. IEEE Trans Circ Syst I: Reg Pap 57(3):725–734
39. Muh-Dey W, Kalim D, Erguvan D, Sheng-Fuh C, Negra R (2012) Investigation of wideband load transformation networks for Class-E switching-mode power amplifiers. IEEE Trans Microw Theory Techn 60(6):1916–1927
40. Apostolidou M, Van der Heijden MP, Leenaerts DMW, Sonsky J, Heringa A, Volokhine I (2009) A 65 nm CMOS 30 dBm Class-E RF power amplifier with 60% PAE and 40% PAE at 16 dB back-off. IEEE J Solid-State Circ 44(5):1372–1379
41. Changkun P, Jeonghu H, Haksun K, Songcheol H (2008) A 1.8-GHz CMOS power amplifier using a dual-primary transformer with improved efficiency in the low power region. IEEE Trans Microw Theory Techn 56(4):782–792
42. Yonghoon S, Sungho L, Cho E, Jaejun L, Sangwook N (2010) A CMOS Class-E power amplifier with voltage stress relief and enhanced efficiency. IEEE Trans Microw Theory Techn 58(2):310–317
43. Brama R, Larcher L, Mazzanti A, Svelto F (2008) A 30.5 dBm 48% PAE CMOS Class-E PA with integrated balun for RF applications. IEEE J Solid-State Circ 43(8):1755–1762
44. Sira D, Thomsen P, Larsen T (2010) Output power control in Class-E power amplifiers. IEEE Microw Wireless Compon Lett 20(4):232–234
45. Dong-Ho L, Changkun P, Jeonghu H, Kim Y, Songcheol H, Chang-Ho L, Laskar J (2008) A load-shared CMOS power amplifier with efficiency boosting at low power mode for polar transmitters. IEEE Trans Microw Theory Techn 56(7):1565–1574
46. Hongtak L, Changkun P, Songcheol H (2009) A quasi-four-pair Class-E CMOS RF power amplifier with an integrated passive device transformer. IEEE Trans Microw Theory Techn 57(4):752–759
47. Godoy PA, SungWon C, Barton TW, Perreault DJ, Dawson JL (2012) A 2.4-GHz, 27-dBm asymmetric multilevel outphasing power amplifier in 65-nm CMOS. IEEE J Solid-State Circ 47(10):2372–2384
48. Reynaert P, Steyaert MSJ (2005) A 1.75-GHz polar modulated CMOS RF power amplifier for GSM-EDGE. IEEE J Solid-State Circ 40(12):2598–2608
49. Walling JS, Taylor SS, Allstot DJ (2009) A Class-G supply modulator and Class-E PA in 130 nm CMOS. IEEE J Solid-State Circ 44(9):2339–2347
50. Singhal N, Pamarti S (2010) A digital envelope combiner for switching power amplifier linearization. IEEE Trans Circ Syst II: Exp Briefs 57(4):270–274

51. Walling JS, Lakdawala H, Palaskas Y, Ravi A, Degani O, Soumyanath K, Allstot DJ (2009) A Class-E PA with pulse-width and pulse-position modulation in 65 nm CMOS. IEEE J Solid-State Circ 44(6):1668–1678
52. Raab FH (2001) Maximum efficiency and output of class-F power amplifiers. IEEE Trans Microw Theory Techn 49(6):1162–1166
53. Huang Min Z, A''ain AKB, Kordesch AV (2007) Two stage integrated Class-F RF power amplifier. In: Int symposium integrated circuits (ISIC 2007), pp 108–110
54. Carls J, Ellinger F, Joerges U, Krcmar M (2009) Highly-efficient CMOS C-band Class-F power amplifier for low supply voltages. IET Electron Lett 45(24):1240–1241
55. Fortes F, do Rosario MJ (2001) A second harmonic Class-F power amplifier in standard CMOS technology. IEEE Trans Microw Theory Techn 49(6):1216–1220
56. Hsien-Yuan L, Jhih-Hong C, Hwann-Kaeo C, Shin-Ming W (2009) Harmonic control network for 2.6 GHz CMOS Class-F power amplifier. In: IEEE international symposium circuits and systems (ISCAS 2009), pp 1321–1324
57. Kuo TC, Lusignan BB (2001) A 1.5 W class-F RF power amplifier in 0.2 mm CMOS technology. In: IEEE international digital technology papers solid-state circuits conference (ISSCC 2001), pp 154–155, 442
58. Shirvani A, Su DK, Wooley BA (2002) A CMOS RF power amplifier with parallel amplification for efficient power control. IEEE J Solid-State Circ 37(6):684–693
59. Hamedi-Hagh S, Salama CAT (2004) CMOS wireless phase-shifted transmitter. IEEE J Solid-State Circ 39(8):1241–1252

Chapter 4
CMOS Performance Issues

Abstract This chapter details how CMOS processes limit the performance of fully integrated linear PAs. The low V_{DD} of modern, submicron CMOS processes, along with breakdown phenomena and hot carrier degradation, are the main limitations. On top of that, these limitations are also accompanied by other effects such as reduced output voltage headroom due to the V_{KNEE}, the low quality factor of integrated inductor and transformers, transistor parasitics, substrate losses or stability issues. All these limitations have a direct impact on the linearity, the output power levels and the efficiency of PAs.

Low Supply Voltage

The downscaling of CMOS technology in order to improve the performance of CMOS processes requires the reduction of the channel length (L_g) and the V_{DD} so that the transistor size and the power consumption are reduced along with the improvement of the process speed. Fig. 4.1 shows the predicted evolution of the V_{DD} and L_g for three different CMOS technologies in the years to come, demonstrating that channel length and supply voltage reduction is an unavoidable fact in nm CMOS processes [1].

Although this evolution in CMOS technology is undoubtedly beneficial for digital circuits, it is not the case for analog devices and it is highly detrimental for PAs in particular because, at shown in Chap. 3, the linear output power capability of a PA is directly related to the value of the V_{DD} that the PA can deal with.

Effect of the Knee Voltage

As introduced in Chap. 3, the V_{KNEE} parameter is extremely important in PAs because it imposes an extra limitation on the drain voltage swing and therefore on the maximum linear output power [2]. In submicron CMOS processes its effects

H. Solar Ruiz and R. Berenguer Pérez, *Linear CMOS RF Power Amplifiers*,
DOI: 10.1007/978-1-4614-8657-2_4,

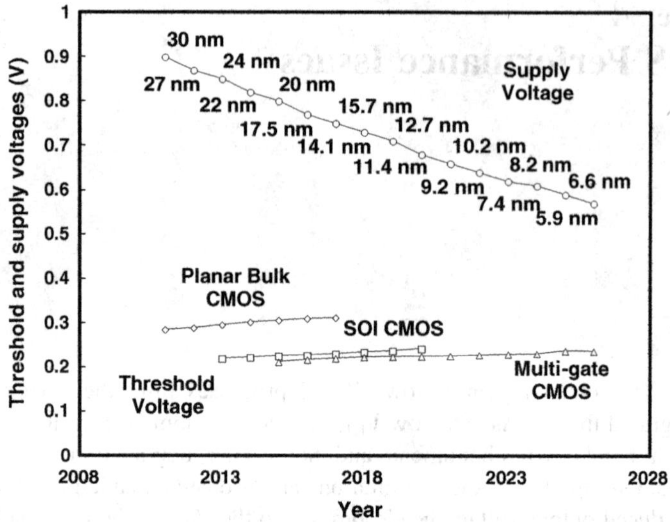

Fig. 4.1 Predicted evolution of the supply voltage and threshold voltage for different CMOS technologies. *Source* ITRS 2011

must be taken into account as it may constitute a considerable percentage of V_{DD}. The effect of the V_{KNEE} parameter was included in an idealized V_{DS}–I_{DS} model in Chap. 3 as a first approximation. However, as shown in Fig. 4.2, the V_{KNEE} is not so easily quantified in a more realistic V_{DS}–I_{DS} model but its effect is also the same; it reduces the maximum drain voltage headroom because the voltage should not swing down to zero.

As shown in the V_{DS}–I_{DS} curves in Fig. 4.2, the V_{KNEE} can be identified with the saturation drain voltage parameter (V_{DSAT}). The V_{DSAT} is the minimum V_{DS} at which the transistor is still in saturation and thus V_{DSAT} is equal to V_{GS}–V_{TH} [3].

When the drain voltage swings down to V_{KNEE} the transistor enters the conductive region, which means that the I_{DS} drops and the maximum linear output power is consequently reduced. There is a way to recover some of the PA performance by reducing the drain voltage swings so that the transistor does not enter the conductive region and the drain current swings are not affected [2]. However, the performance of the PA in terms of linear output power will be always degraded due to the reduced voltage swing from V_{DD} to V_{DD}–V_{KNEE}. As already shown in Chap. 3, if V_{KNEE} is considered the maximum linear output power is (4.1).

$$P_{OUT} = \frac{(V_{DD} - V_{KNEE})^2}{2R_{OPT}} \tag{4.1}$$

The effect of V_{KNEE} on the P_{OUT} can be quantified based on (4.1). If a 0.5 Watt-level 1 V V_{DD} PA is chosen, the linear output power reduction can be observed in Fig. 4.3 in terms of the percentage of V_{KNEE} with respect to V_{DD}.

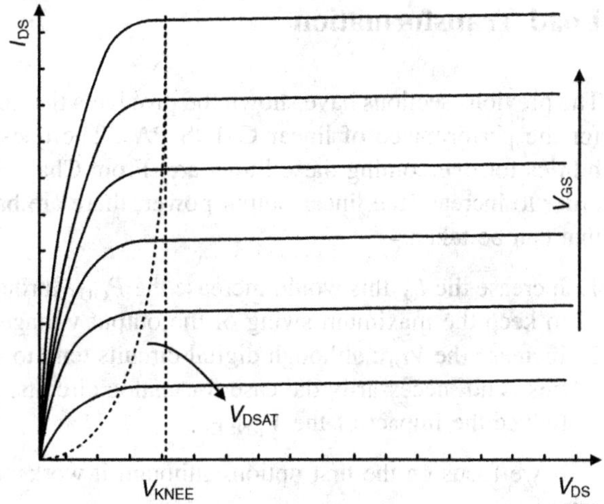

Fig. 4.2 The V_{KNEE} and V_{DSAT} parameters in the V_{DS}–I_{DS} model of a transistor

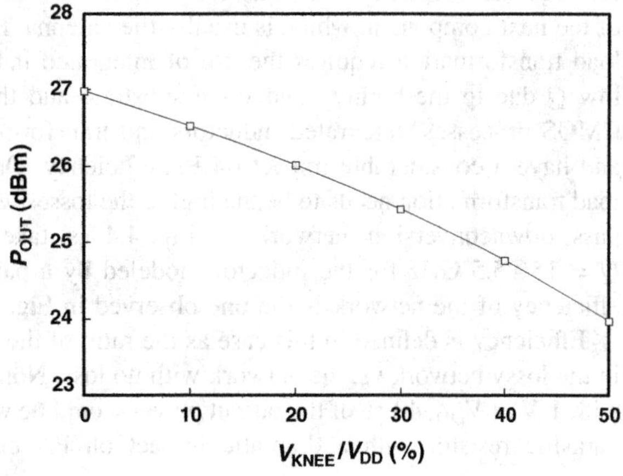

Fig. 4.3 Effect of the V_{KNEE} parameter in the maximum linear output power levels

It can be seen that the maximum linear output power is halved if the V_{KNEE} reaches 50 % of the V_{DD}. Although a V_{KNEE} being 50 % of the V_{DD} may sound exaggerated, the trend in technology does not say so [4]. Therefore, the scaling-down of the CMOS technology does not only impose the problem of low V_{DD} but also that of limited output voltage headroom due to relatively high V_{KNEE}. The effect of the V_{KNEE} must always be taken into consideration in PA design. In fact, the model presented in Chap. 6 introduces this effect in order to have a more realistic calculation of linear CMOS PA performance.

Load Transformation

The previous sections have shown the problems that low V_{DD} and high V_{KNEE} pose for the performance of linear CMOS PAs. The question now is what the possibilities for overcoming these limits are. From Chap. 3 it can be concluded that in order to increase the linear output power, there are basically only two main steps that can be taken:

1. Increase the I_Q: this would increase the P_{OUT} through R_{LOAD} reduction in order to keep the maximum swing of the output voltage constant.
2. Increase the V_{DD}: although digital circuits tend to use the lowest V_{DD} possible, this is not necessarily the case for analog circuits. Increasing the V_{DD} will also reduce the impact of the V_{KNEE}.

If we focus on the first option, although it works, several negative side effects arise.

Increasing the I_Q requires aggressive load transformation. In the aforementioned example of a 0.5 Watt-level PA with a V_{DD} of 1 V, even if the V_{KNEE} effect is not considered, a 50:1 load transformation ratio would be required from the 50 Ω of the next component, which is usually the antenna. For fully integrated PAs, any load transformation requires the use of integrated inductors or transformers with low Q due to the highly conductive substrate and thin metal tracks in standard CMOS processes. Integrated inductors and transformers are therefore very lossy and have a considerable impact on PA efficiency. On top of that, the higher the load transformation needs to be the higher the losses will be [5]. If the simple high-pass downconversion network in Fig. 4.4 is taken as an example, with a $Q = 15@5.5$ GHz for the inductor modeled by a parasitic series resistor r_P, the efficiency of the network is the one observed in Fig. 4.5.

Efficiency is defined in this case as the ratio of the power dissipated at the load in the lossy network vs. the network with no loss. Note that in a 0.5 Watt-level PA with 1 V of V_{DD}, 40 % of the output power would be wasted on the output network parasitic resistor with a dramatic impact on PA efficiency. Again, the model

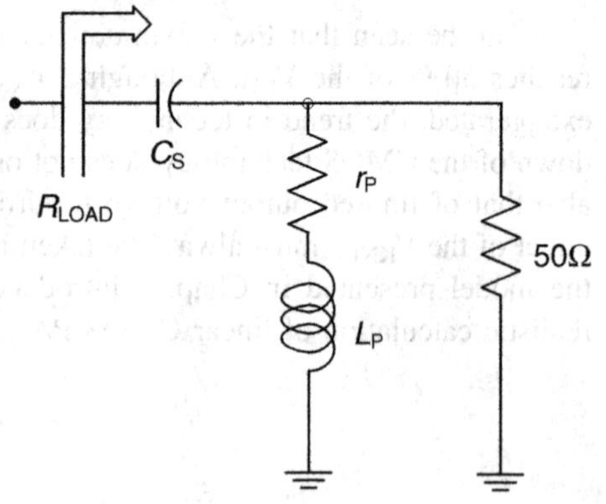

Fig. 4.4 Schematic view of the high-pass downconversion network

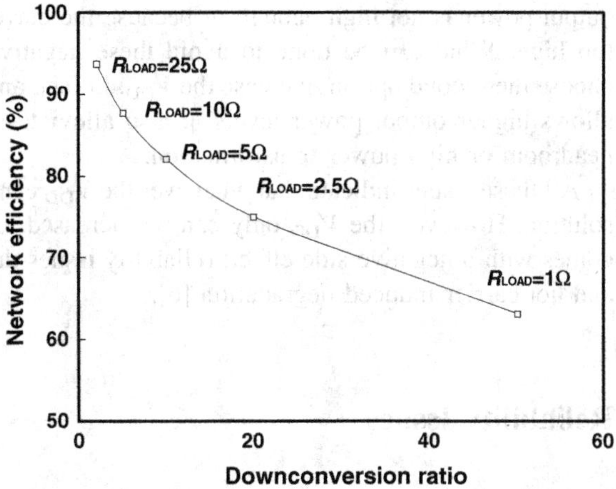

Fig. 4.5 Efficiency of the high-pass downconversion network versus the downconversion ratio

presented in Chap. 6 will quantify the impact of this effect on PA efficiency and the output power.

Another side effect comes from the fact that high current levels flowing through limited conductive metal tracks cause noticeable voltage drops. It must be borne in mind that metal tracks with limited conductivity and low Q drain inductors are placed in the DC current path in fully integrated PA. As the inductor shows no negligible parasitic resistance, the actual V_{DD} gets reduced. A parasitic DC resistor of 1 Ω would spoil the performance of a 0.5-Watt level PA with a V_{DD} of 1 V because the theoretical voltage drop would be 0.5 V!

In addition, any integrated process limits the maximum current density flowing through a metal track. A value of a few units of mA/µm is given to the different metal layers of any CMOS process along with a limit for the maximum metal track width. Therefore, if higher currents are necessary, the need for wider metal will increase the parasitic capacitance and the losses. A way to reduce the metal track width is to connect several layers in parallel by using vias, which also solves the problem of the maximum metal track width. However, the distance from the resulting metal track to the substrate decreases, which counteracts the parasitic reduction.

Finally, high currents cause problems for the transistors that must deal with them. The CMOS process sets a limit to the maximum current level a transistor must withstand through its channel. This is not a crucial problem because transistor width is usually high and an interdigital topology with a large number of fingers is employed. However, high current levels force higher power dissipation to the device. This causes a rise in the temperature and a corresponding worsening of the process performance. At high temperatures carrier mobility decreases, and the transistor either shows a higher threshold voltage or the maximum current density of the metal tracks is reduced.

The final consequence of excessively high current levels is that the efficiency of the PA would be unacceptably low for the final application, either because the output power is not high enough or because the current or power consumption is too high. What can be done to avoid these negative effects? The answer is to choose the second option: increase the V_{DD}. In fact, an increase in the V_{DD} not only allows higher output power levels, it also alleviates the problem of low voltage headroom or high power transformation.

All these issues indicate that whenever the V_{DD} can be increased, that is the best solution. However, the V_{DD} only can be increased to a limited extent because it comes with a negative side effect: reliability issues due to breakdown phenomena and hot carrier induced degradation [6].

Reliability Issues

Reliability issues come from the fact that the downscaling of the L_g and V_{DD} must be followed by a subsequent reduction in oxide thickness (t_{ox}) in order to control the short channel effects. This is observed in the predicted evolution for three different CMOS technologies given in Fig. 4.6.

As t_{ox} is scaled down, the electrical fields in the oxide layer increase and consequently the performance of the submicron transistors is compromised due to reliability issues. These issues are caused by different phenomena that place in the oxide layer and the channel of the CMOS transistors. The ones that are most critical for PAs are gate oxide breakdown and hot carrier degradation.

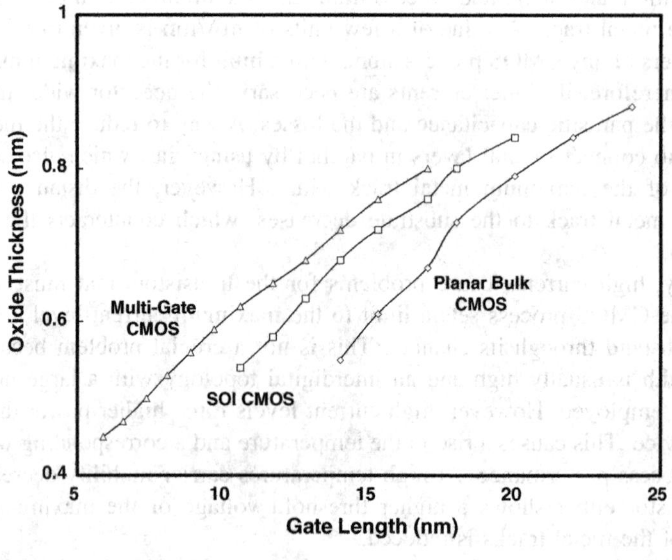

Fig. 4.6 Predicted evolution of oxide thickness versus the gate length in CMOS technologies. *Source* ITRS 2011

Oxide Breakdown

Oxide breakdown phenomena take place when carriers (electrons in NMOS and holes in PMOS) in the transistor channel gain sufficient energy to tunnel through the energy barrier between the silicon and the oxide under the influence of high electrical fields. Oxide breakdown events are typically classified as soft breakdown or hard breakdown, depending on the consequences in transistor performance.

As observed in Fig. 4.7, oxide breakdown begins when the electric field across the gate dielectric causes the generation of electrical defects or traps in the oxide layer. In the first stages, the density of the traps is still low, so the traps are isolated. However, as the density of the traps in the oxide layer increases, they start to overlap so that a conduction path is created and breakdown occurs [7]. In such a situation, the leakage current of the device increases noticeably with respect to the defect-free device. This type of breakdown is defined as soft breakdown because only few conduction paths still exist in specific spots. For modern CMOS processes in which ultrathin oxides and lower voltages are employed, the soft breakdown is more frequently observed than hard breakdown [8]. Soft breakdown phenomena may not destroy the functionality of a device, but it causes strong performance degradation [9–11].

In the next stage, as shown in Fig. 4.7, these conduction paths allow the generation of new traps due to the increase in heat and layer conductance so that the breakdown spot is laterally propagated [12, 13]. The final stage is that the silicon within the conduction path melts, releasing oxygen so that a silicon path is created. This situation is defined as hard breakdown of the oxide layer [12]. After a hard breakdown, the transistor is nonfunctional as a result of the damage from the gate oxide [14].

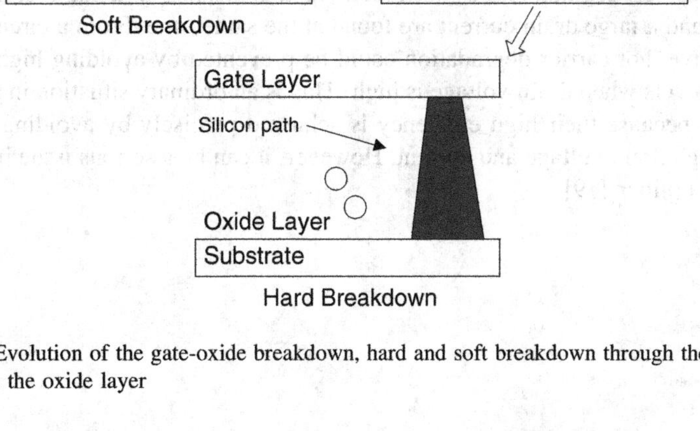

Fig. 4.7 Evolution of the gate-oxide breakdown, hard and soft breakdown through the creation of traps in the oxide layer

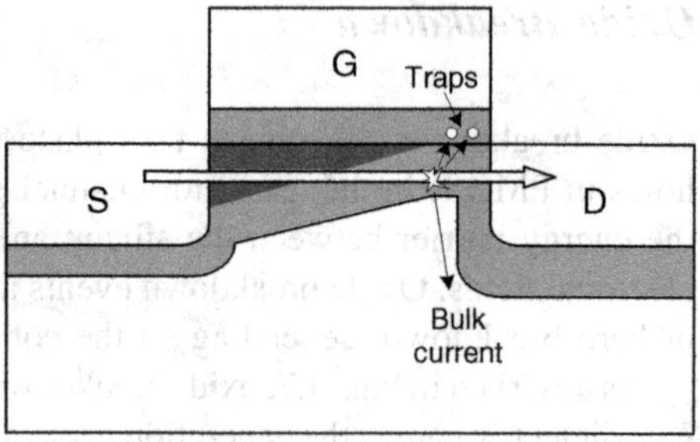

Fig. 4.8 Hot carrier degradation in a MOS transistor

Hot Carrier Degradation

This phenomenon is another manifestation of degradation in transistor characteristics, but whereas the breakdown phenomena are seen when there is no potential difference between the source and the drain, hot carrier stress take place under the influence of high lateral fields in the channel. As observed in Fig. 4.8, carriers traveling through the channel can gain sufficient kinetic energy and create an electron–hole pair through impact ionization within the channel. The holes enter the substrate and lead to the substrate current, while the electrons enter the gate oxide [15].

The hot carrier phenomenon degrades charge mobility and consequently the V_{TH} and g_m characteristics of the transistor are also affected [9, 16]. Therefore, in MOS transistors, the performance degradation due to hot carrier stress can be examined via device parameters such as transconductance degradation, threshold voltage increase or mobility shift [10].

Transistor performance degradation due to hot carrier is gradual over a period of time and becomes evident during the first few hours of continuous operation [6, 17]. It is not a catastrophic phenomenon, but as hot electrons enter the gate, they may create traps so that they contribute to the oxide breakdown phenomenon. In fact, the existence of channel hot carrier decreases the time to CMOS transistor breakdown [18].

Hot carrier degradation is noticeable in a transistor if a high drain-source voltage and a large drain current are found at the same time. From a circuit design perspective, hot carrier degradation could be prevented by avoiding high channel current levels when drain voltage is high. This is an ordinary situation in a switch-type PA because their high efficiency is achieved precisely by avoiding simultaneous high drain voltage and current. However, it can be a serious issue in a linear power amplifier [19].

Reliability Projection

The reliability of an integrated process is the foundry's guarantee that the performance of the CMOS process remains within specification for a determined period of time. From the point of view of oxide breakdown, the usual definition adopted by the microelectronics industry is that after a certain amount of time (years) of operation at the nominal conditions only a small number of devices (e.g. 100 ppm) can be broken [20].

Although the definition is clear, the main difficulties come from the way these data are obtained. Logically, it is not possible to test a large number of transistors for several years, so some kind of accelerated stress condition tests must be performed. Multiple tests and techniques exist for the evaluation of CMOS reliability, but the time dependent dielectric breakdown (TDDB) test is the standard methodology for developing operating lifetime reliability projections [21]. The TDDB is then a measure of the oxide breakdown that occurs over time due to trap generation at normal voltage but based on stress conditions for the transistor until breakdown takes place. The TDDB tests are described in statistical terms because the breakdown phenomena are also statistical: two identical devices subjected to the same stress may break down at different times. That is why for a specific CMOS process it is possible to define only a failure probability as a function of time.

Based on this type of test, an extrapolation curve is generated in order to project the reliability of devices under real operating conditions. At the present time there is no agreement regarding the correct oxide-reliability extrapolation in modern CMOS processes though several physical mechanisms, e.g. the percolation model, and computational models, e.g. the anode hole-injection model (AHI), have been developed in order to obtain accurate results [14].

The process of projecting the reliability of a CMOS process is then based on several relatively short-term accelerated tests, such as the TDDB, to which failure mechanisms and breakdown projection models are applied. This procedure predicts a device lifetime based on the magnitude of the electric field across the gate oxide or the gate voltage (usually depending of the thickness of the gate oxide layers [22]), as shown in Fig. 4.9.

The reliability test for transistors is transferred to transistor models for simulations. The usual parameter that sets a limit concerning device reliability is the oxide breakdown voltage parameter (vbox), which is numerically related to the transistor oxide thickness parameter by means of the maximum field across the oxide layer. A typical relation between vbox and t_{ox} in the Spectre BSIM transistor model is expressed in (4.2), which in the end reflects that the reliability of the transistors is ensured for a field strength of less than 10 MV/cm.

$$vbox(V) = 1 \cdot 10^9 * t_{ox}(nm) \tag{4.2}$$

It must be noted that the field across the bottom of the gate oxide is not constant along the source to drain region, so the device stress is a function of position, as it can

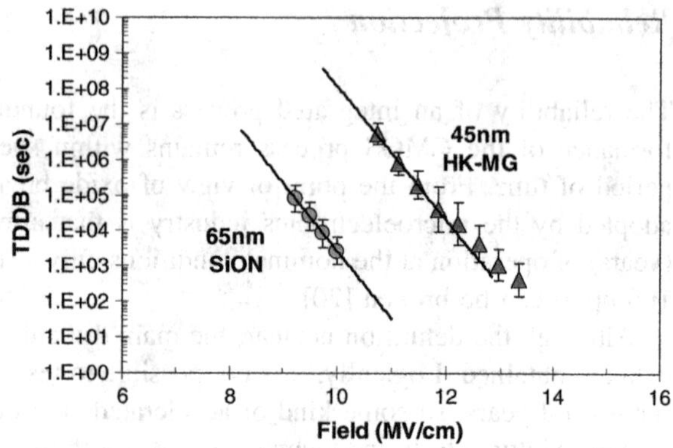

Fig. 4.9 Time dependent dielectric breakdown (TDDB) test results in a 45 nm CMOS process. *Source* Intel® Technology Journal

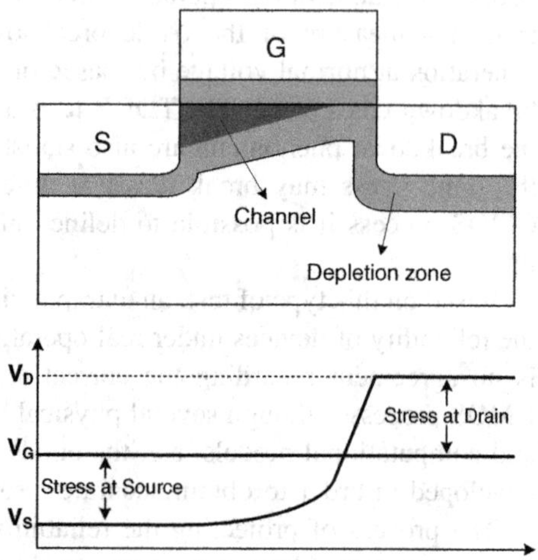

Fig. 4.10 Voltage distribution in the channel of a MOS transistor

be observed in Fig. 4.10. The highest stress areas occur at the source and drain oxide edges. Therefore, in PA design the gate-to-source (V_{GS}) and the gate-to-drain (V_{GD}) voltages what the designer should focus on in order not to exceed vbox [19].

Reliability Under RF Stress

The aforementioned models predict CMOS process reliability by means of the DC gate voltage values, but transistors in PAs suffer from RF voltage stress. In general, PA transistors are biased with voltages below the breakdown limits in order to ensure reliability, but RF peaks usually exceed the breakdown voltage limits of the technology. Several studies indicate that CMOS reliability is preserved while allowing operating voltages that exceed the breakdown limit at high frequencies

[11, 23], and the crucial parameter is the rms level of the applied voltage instead of the maximum values of RF peaks [11].

On the other hand, what is important to a PA designer is predicting the impact of degradation not for the PA transistors but primarily for the overall PA performance. In terms of RF performance degradation, the soft breakdown and the hot carrier degradation are the most important issues. There are several studies in the literature that quantify the effects of soft breakdown and the hot carrier injection on different RF circuits [9, 10, 24–26]. As mentioned previously, results show that PAs may suffer degradation during the initial stress stages and that the performance of the PA is gradually stabilized. Even the previous performance can be recovered by modifying the transistor gate bias [6].

The rule of thumb, then, in the design of reliable PAs is to bias transistors with DC voltages and allow a certain RF voltage swing overdrive that may cause certain degradation that is stabilized over time. The RF overdrive must be studied for the specific PA and CMOS process so that the stabilization of the performance is observed. This design procedure should be also combined with techniques that allow the voltage stress to be relaxed, such as the use of cascode transistors or differential topologies [9, 27, 28].

Transistor Parasitics

The fact that CMOS transistors require low supply voltage for reliable performance along with their relatively low transconductance forces the transistor size of the PA output to be increased. However, large transistors are accompanied by high transistor parasitics which degrades the overall performance of the PA in several ways.

First, large output transistors present large input capacitance and thus an increase in the loading requirements for the driving stage. This results in higher power consumption of the driver stage, and consequently reduced efficiency [29, 30].

In addition, the linearity of the PA is also degraded because PA linearity is highly affected by the transistor parasitic capacitances that are the cause of the AM-PM distortions [31].

Finally, the stability of the PA is compromised due to the feedback paths provided by the parasitics. For example, the increase of the drain-to-gate capacitance of the output transistor is a main source of instability because it reduces the unilateral behavior of the transistor [31, 32].

Substrate Issues

Standard CMOS processes are fabricated on a bulk silicon substrate with a doping level that drives the substrate resistance to a value of about 10 Ω-cm [28, 33]. The conductive nature of the substrate is a primary reason for the losses on both passive and active devices, and in particularly for the low quality factor of passive components, mainly the planar inductors and transformers fabricated on top of the substrate. The relatively high capacitive coupling effect (given the short distance to the substrate) and the current losses induced in the substrate (such as the eddy currents and conductive or displacement currents) are behind the low quality factor values of planar inductors and transformers [34]. The losses in silicon can be reduced by using ground-shielding techniques between the inductor and the silicon substrate. Ground shielding may lead to an increase in Q, but in this case at the expense of a reduction of the self-resonant frequency which can be an important issue if wide track inductors are required. In general, the losses in the silicon substrate of standard CMOS processes cannot be eliminated without any other degradation of the inductor characteristics [35].

Another issue related to the use of conductive substrate concerns to stability. Conductive substrates may create feedback paths between the input and the output of amplifying stages. This effect must be considered and measures such as stage isolation should be taken. The use of substrate contacts on the surface of the silicon substrate between the stages is a proper measure as these contacts provide a proper return path to ground for potential substrate currents flowing through the substrate [34].

Stability Issues

Stability is a major concern for any type of RF and microwave amplifier. In the case of PAs this problem is exacerbated, as PAs may exhibit several types of instabilities simultaneously. In addition to linear feedback mechanisms, which are a cause of oscillation in any amplifier, strong nonlinearity conditions in PAs are also a source of instabilities [36, 37].

In general, PA instabilities can be divided into two main sources. The first source comes from the instabilities caused by the transistor capacitance parasitics, combined with parasitic circuit inductances. For example, the gate to drain feedback capacitor in conjunction with the drain inductor is a source of potential instability [31, 32, 36]. The second source is the instability created by the combination of bypass capacitors, along with the parasitic ground and supply inductances of the bondwire or supply line connections [36].

In general, these problems can be solved by adding a series resistor to the loop causing the instability, thus reducing the quality factor of this loop. However,

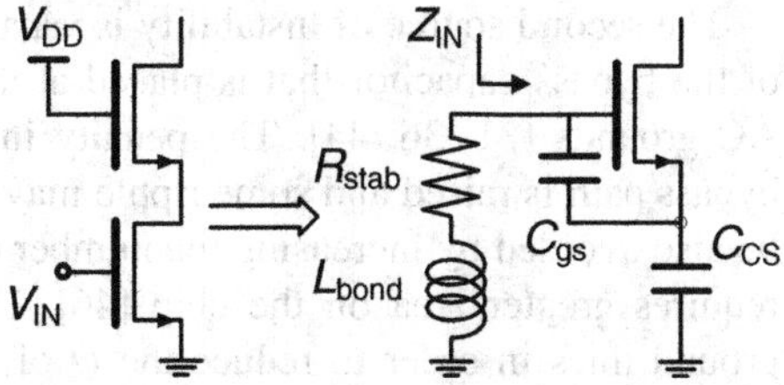

Fig. 4.11 Cascode stage instability issues

adding a resistor may cause performance degradation of the PA so that using the proper stabilization mechanism is crucial in order to minimize that degradation.

In addition, potential unstable sources are multiplied in the case of multistage PAs given the increase of loops that may create the instabilities. For example, an extra potential instability can be caused by the input capacitance of the transistor of the output stage combined with the drain inductor of the driver stage.

The aforementioned first source of instability can be solved by using RC feedback networks between the drain and the gate of the transistor, although this produces the negative effect of gain reduction [38–42].

With this type of instability the use of the cascode transistor is also of help because it makes the stage more unilateral as it isolates the output from the input of the transistor, and hence the stage is more stable [33, 43]. In this way, the use of the cascode transistor not only allows the PA to handle greater output power levels but it also improves PA stability. However, the cascode transistor itself can be also a source of instability due to the insufficient AC grounding of the common-gate terminal of the cascode transistor [31, 33]. If the parasitic inductance between the gate and ground of the cascode transistor is high enough, it can force the PA to oscillate due to the inductor capacitance parasitics, as can be observed in Fig. 4.11.

The input impedance at the gate of the cascode device (ignoring the gate to drain parasitic capacitances) is (4.3).

$$Z_{IN} = \frac{1}{j\omega C_{gs}} + \frac{1}{j\omega C_{CS}} - \frac{g_m}{\omega^2 C_{gs} C_{CS}} \tag{4.3}$$

Where C_{CS} is the equivalent output inductance of the common-source transistor and C_{gs} is the gate-to-source inductance of the common-gate transistor. As (4.3) shows, the input impedance real part has a negative component. This means that the circuit may oscillate if a parasitic inductance is connected to the gate of the transistor. This potential oscillation can be solved by introducing a series resistance to the gate of the cascode transistor because it reduces the Q of the network. In order to keep the stage stable, the value of the series resistor must be high enough to counteract the negative component of the input impedance real part (4.4).

$$R_{stab} > \frac{g_m}{\omega^2 C_{gs} C_{CS}} \tag{4.4}$$

The second source of instability is usually solved by reducing the quality factor of the bypass capacitor that is placed at the supply lines to allow low impedance AC grounds [31, 36, 44]. The penalty in this case is that the impedance of the bypass path is raised and some ripple may appear [45]. This type of instability can be also avoided by increasing the number of bondwire connections but this in turn requires greater area on the chip [46]. Using series resistors in the supply and ground lines in order to reduce the Q of the instability loop must be avoided as these resistors greatly degrade the performance of the PA.

As mentioned above, instability problems are aggravated in the case of multistage PAs. Due to the parasitic inductance on the ground, an AC voltage (ground bounce) is created. The AC voltage causes degeneration that reduces not only the gain and PAE of the PA but in a multistage amplifier the on-chip ground bounce may also cause undesirable feedback between stages and instability [47]. In order to suppress this local feedback between stages, each stage should be isolated on-chip and be ultimately connected to the off-chip ground plane with separate bondwires. Again, this comes at the cost of greater chip area if many ground pads are required. Another measure to isolate the stages is to use of transformers between stages; this may improve stability at low frequencies due to the high-pass characteristic of the transformers [31, 48]. Special care must be taken, however, in order to avoid positive feedback due to the coupling effect between the output and the input of the PA [49, 50].

The use of a differential configuration is helpful when tackling stability issues. It is true that the differential configuration is not immune to instabilities because common-mode oscillations may take place [36, 51]. However, the symmetries of the differential PA can be exploited in order to create virtual AC grounds in the common-mode connection nodes so that those nodes carry only current at dc and even harmonics [52, 53]. In such a situation, the amplifier performance is desensitized from the ground and supply parasitic inductances [46, 52]. For that reason, in the differential configuration, the resistors can be connected in the common-mode paths at the virtual ground in order to stabilize the circuit but they do not degrade the performance of the differential amplifier at RF [47, 48]. Note that supply and ground separations between stages are only feasible in differential circuits for differential signaling provides a specified signal current return path [51].

Conclusions

This chapter has discussed the main limitations and issues that the CMOS technology imposes on the design of linear CMOS PAs and that the designer must deal with. It is crucial that these limitations be taken into account when designing linear CMOS PAs for a specific communication standard so that the appropriate CMOS process and PA architecture are chosen in order to meet the requirements. Any CMOS PA design can easily fail, resulting in a considerable waste of time and engineering effort whenever any of the CMOS limitations are disregarded. The model discussed in Chap. 6 will be of great help as it covers many of these limitations. Preliminary results of PA performance can be extracted while avoiding long simulation times.

References

1. International Technology Roadmap for Semiconductors (ITRS) (2011) Edition: Process integration, devices, and structures. http://www.itrs.net/Links/2011ITRS/Home2011.htm. Accessed 30 Mar 2013
2. Cripps SC (1999) RF power amplifiers for wireless communications. Artech House, Norwood
3. Razavi B (2001) Design of analog CMOS integrated circuits. McGraw Hill, New York
4. Kuo TC, Lusignan BB (2001) A 1.5 W class-F RF power amplifier in 0.2 mm CMOS technology. In: IEEE international digest technical papers solid-state circuits conference (ISSCC 2001), pp 154–155
5. Aoki I, Kee SD, Rutledge DB, Hajimiri A (2002) Distributed active transformer-a new power-combining and impedance-transformation technique. IEEE Trans Microw Theory Techn 50(1):316–331
6. Vathulya VR, Sowlati T, Leenaerts D (2001) Class 1 bluetooth power amplifier with 24 dBm output power and 48 % PAE at 2.4 GHz in 0.25 mm CMOS. In: Proceedings of european solid-state Circuits Conference (ESSCIRC 2001), pp 57–60
7. Degraeve R, Groeseneken G, Bellens R, Depas M, Maes HE (1995) A consistent model for the thickness dependence of intrinsic breakdown in ultra-thin oxides. In: IEEE international electron devices meeting technical digest (IEDM 1995), pp 863–866
8. Weir BE, Silverman PJ, Monroe D, Krisch KS, Alam MA, Alers GB, Sorsch TW, Timp GL, Baumann F, Liu CT, Ma Y, Hwang D (1997) Ultra-thin gate dielectrics: they break down, but do they fail? In: IEEE international electron devices meeting technical digest (IEDM 1997), pp 73–76
9. Enjun X, Yuan JS, Hong Y (2004) CMOS RF and DC reliability subject to hot carrier stress and oxide soft breakdown. IEEE Trans Device Mater Rel 4(1):92–98
10. Li Q, Zhang J, Li W, Yuan JS, Chen P, Oates AS (2001) RF circuit performance degradation due to soft breakdown and hot carrier effect in deep-submicrometer CMOS technology. IEEE Trans Microw Theory Tech 49(9):1546–1551
11. Larcher L, Sanzogni D, Brama R, Mazzanti A, Svelto F (2006) Oxide breakdown after RF stress: experimental analysis and effects on power amplifier operation. In: 44th Annual IEEE international reliability physics symposium proceedings, pp 283–288
12. Lin HC, Lee DY, Lee CY, Chao TS, Huang TY, Wang T (2001) New insights into breakdown modes and their evolution in ultra-thin gate oxide. In: International symposium on VLSI technology, systems, and applications, procedings of technical paper, pp 37–40
13. Depas M, Nigam T, Heyns MM (1996) Soft breakdown of ultra-thin gate oxide layers. IEEE Trans Electron Devices 43(9):1499–1504
14. Stathis JH (2002) Reliability limits for the gate insulator in cmos technology. IBM J Res Dev 46(2/3):265–286
15. Cheng TW (1992) Hot carrier design considerations for MOS devices and circuits. Springer, New York
16. Takeda E, Yang CY, Miura-Hamada (1995) Hot carrier effects in MOS devices. Academic Press, New York
17. Sowlati T, Leenaerts DMW (2003) A 2.4-GHz 0.18 mm CMOS self-biased cascode power amplifier. IEEE J Solid-State Circuits 38(8):1318–1324

18. Crupi F, Kaczer B, Groeseneken G, De Keersgieter A (2003) New insights into the relation between channel hot carrier degradation and oxide breakdown short channel nMOSFETs. IEEE Electron Device Lett 24(4):278–280
19. Aoki I, Kee S, Magoon R, Aparicio R, Bohn F, Zachan J, Hatcher G, McClymont D, Hajimiri A (2008) A fully-integrated quad-band GSM/GPRS CMOS power amplifier. IEEE J Solid-State Circuits 43(12):2747–2758
20. Dabrowski J, Weber ER (2004) Predictive simulation of semiconductor processing: status and challenges. Springer, Berlin
21. Hicks J, Bergstrom D, Hattendorf M, Jopling J, Maiz J, Pae S, Prasad C, Wiedemer J (2008) 45 nm Transistor reliability. Intel Tech J 12(2):131–144
22. Wu EY, Vayshenker A, Nowak E, Sune J, Vollertsen RP, Lai W, Harmon D (2002) Experimental evidence of T_{BD} power-law for voltage dependence of oxide breakdown in ultrathin gate oxides. IEEE Trans Electron Devices 49(12):2244–2253
23. Sasse GT, Kuper FG, Schmitz J (2008) MOSFET degradation under RF stress. IEEE Trans Electron Devices 55(11):3167–3174
24. Enjun X (2005) Hot carrier effect on CMOS RF amplifiers. In: 43th Annual IEEE international reliability physics symposium Proceedings, pp 680–681
25. Naseh S, Deen MJ, Marinov O (2002) Effects of hot-carrier stress on the RF performance of 0.18 μm technology NMOSFETs and circuits. In: 40th Annual IEEE international reliability physics symposium Proceedings, pp 98–104
26. Qiang L, Wei L, Jinlong Z, Yuan JS (2002) Soft breakdown and hot carrier reliability of CMOS RF mixer and redesign. IEEE radio frequency integrated circuits symposium (RFIC 2002), 399–402
27. Hyoung-Seok O, Cheon-Soo K, Yu H, Kim C, x, Ki (2006) A fully-integrated +23-dBm CMOS triple cascode linear power amplifier with inner-parallel power control scheme. In: IEEE radio frequency integrated circuits symposium (RFIC 2006)
28. Niknejad AM, Chowdhury D, Jiashu C (2012) Design of CMOS power amplifiers. IEEE Trans Microw Theory Techn 60(6):1784–1796
29. Reynaert P, Steyaert M (2006) RF power amplifiers for mobile communications. Springer, The Netherlands
30. Lee Hongtak, Park Changkun, Hong Songcheol (2009) A Quasi-Four-Pair Class-E CMOS RF power amplifier with an integrated passive device transformer. IEEE Trans Microw Theory Techn 57(4):752–759
31. Onizuka K, Ishihara H, Hosoya M, Saigusa S, Watanabe O, Otaka S (2012) A 1.9 GHz CMOS power amplifier with embedded linearizer to compensate AM-PM distortion. IEEE J Solid-State Circuits 47(8):1820–1827
32. Deferm N, Osorio JF, Anton de Graauw, Reynaert P (2011) A 94 GHz differential power amplifier in 45 nm LP CMOS. In: IEEE radio frequency integrated circuits symposium (RFIC 2011), pp 1–4
33. Komijani A, Natarajan A, Hajimiri A (2005) A 24-GHz, +14.5-dBm fully integrated power amplifier in 0.18-μm CMOS. IEEE J Solid-State Circuits 40(9):1901–1908
34. Niknejad AM, Meyer RG (2001) Analysis of eddy-current losses over conductive substrates with applications to monolithic inductors and transformers. IEEE Trans Microw Theory Techn 49(1):166–176
35. Burghartz JN (1998) Progress in RF inductors on silicon-understanding substrate losses. In: IEEE international electron devices meeting technical digest (IEDM 1998), 523–526
36. Chowdhury D, Reynaert P, Niknejad AM (2008) Transformer-coupled power amplifier stability and power back-off analysis. IEEE Trans Circuits Syst II: Exp Briefs 55(6):507–511
37. Albulet M (2001) RF power amplifiers. Noble Publishing, Atlanta
38. Kim Jihwan, Kim Woonyun, Jeon Hamhee, Huang Yan-Yu, Yoon Youngchang, Kim Hyungwook, Lee Chang-Ho, Kornegay KT (2012) A fully-integrated high-power linear CMOS power amplifier with a parallel-series combining transformer. IEEE J Solid-State Circuits 47(3):599–614

39. Po-Chih Wang, Chia-Jun Chang, Wei-Ming Chiu, Pei-Ju Chiu, Chun-Cheng Wang, Chao-Hua Lu, Kai-Te Chen, Ming-Chong Huang, Yi-Ming Chang, Shih-Min Lin, Ka-Un Chan, Ying-His Lin, Lee C (2007) A 2.4 GHz Fully Integrated Transmitter Front End with +26.5-dBm On-Chip CMOS Power Amplifier. In: IEEE Radio Frequency Integrated Circuits Symposium (RFIC 2007), pp 263–266
40. Hyunji Koo, Bonhoon Koo, Songcheol Hong (2012) Highly efficient 24-GHz CMOS linear power amplifier with an adaptive bias circuit. In: Asia-Pacific Microwave Conference Proceedings(APMC 2012), pp 7–9
41. Dal Fabbro PA, Meinen C, Kayal M, Kobayashi K, Watanabe Y (2006) A dynamic supply CMOS RF power amplifier for 2.4 GHz and 5.2 GHz frequency bands. In: IEEE Radio Frequency Integrated Circuits Symposium (RFIC 2006), pp 4
42. Murad SAZ, Pokharel RK, Kanaya H, Yoshida K (2009) A 3.1 - 4.8 GHz CMOS UWB Power Amplifier Using Current Reused Technique. In: IEEE international conference on wireless communications, networking and mobile computing (WiCom 2009), pp 1–4
43. Xu Z, Gu QJ, Chang M-F (2011) A 100–117 GHz W-band CMOS power amplifier with on-chip adaptive biasing. IEEE Microw Wireless Compon Lett 21(10):547–549
44. Francois B, Reynaert P (2012) A fully integrated Watt-level linear 900-MHz CMOS RF power amplifier for LTE-applications. IEEE Trans Microw Theory Techn 60(6):1878–1885
45. Chowdhury D, Hull CD, Degani OB, Yanjie Wang, Niknejad AM (2009) A fully integrated dual-mode highly linear 2.4 GHz CMOS power amplifier for 4G WiMax applications. IEEE J Solid-State Circuits 44(12):3393–3402
46. Yan T, Liao H, Li C, Huang R (2007) A 2-GHz fully-differential CMOS power amplifier with virtual grounds to suppress ground bounce. Microw Opt Techn Lett 49(11):2780–2784
47. Cheung TSD, Long JR (2005) A 21-26-GHz SiGe bipolar power amplifier MMIC. IEEE J Solid-State Circuits 40(12):2583–2597
48. Liu JY, Tang A, Ning-Yi Wang, Gu QJ, Berenguer R, Hsieh-Hung Hsieh, Po-Yi Wu, Chewnpu Jou, Chang M-F (2011) A V-band self-healing power amplifier with adaptive feedback bias control in 65 nm CMOS. In: IEEE Radio Frequency Integrated Circuits Symposium (RFIC 2011), pp 1–4
49. Kim Seungwoo, Lee Kyungho, Lee Jongwoo, Kim B, Kee SD, Aoki I, Rutledge DB (2005) An optimized design of distributed active transformer. IEEE Trans Microw Theory Techn 53(1):380–388
50. Lee Hongtak, Park Changkun, Hong Songcheol (2009) A Quasi-Four-Pair Class-E CMOS RF power amplifier with an integrated passive device transformer. IEEE Trans Microw Theory Techn 57(4):752–759
51. Zhao D, He Y, Li L, Joos D, Philibert W, Reynaert P (2011) A 60 GHz 14 dBm power amplifier with a transformer-based power combiner in 65 nm CMOS. Int J Microw Wireless Techn 3(2):99
52. Aoki I, Kee SD, Rutledge DB, Hajimiri A (2002) Fully integrated CMOS power amplifier design using the distributed active-transformer architecture. IEEE J Solid-State Circuits 37(3):371–383
53. Pfeiffer UR, Goren D (2007) A 20 dBm fully-integrated 60 GHz SiGe power amplifier with automatic level control. IEEE J Solid-State Circuits 42(7):1455–1463

Chapter 5
Enhancement Techniques for CMOS Linear PAs

Abstract This chapter describes different techniques that are of interest for boosting the performance of fully integrated CMOS linear PAs. Not only have these techniques been proved to boost linear PA performance, but they can also be fully integrated into a single chip. From the simple use of the cascode transistor or the differential topology to more complex techniques such as the Doherty PA, predistortion, supply modulation or PA combination, these topologies allow the performance of a single PA to be improved in terms of efficiency, linearity or increased output power levels without jeopardizing the full integration of the final solution.

Cascode PAs

As mentioned in Chap. 4, the reliability issues of the CMOS transistor play an important role in the output power levels and the efficiency of the PA. For high efficiency, the PA output voltage may swing up to nearly $2V_{DD}$ and therefore the supply voltage must be limited in order to avoid stressing the CMOS device. However, as also discussed in Chap. 4, the possibility of using a higher supply voltage has several advantages in terms of efficiency, as we avoid excessive current levels and load transformation.

One way to relax the voltage stress of the CMOS transistor that acts as a common-source (CS) in the output stage of the PA is to stack a common-gate (CG) or cascode transistor on top of the CS transistor, as illustrated in Fig. 5.1. The cascode transistor limits the V_{DS} excursions of the CS because CS transistor V_{DS} voltage must be below V_{DD}–V_{TH} so as to keep the CG from entering cut-off.

As mentioned in Chap. 4, when dealing with oxide breakdown, the V_{GD} and V_{GS} voltages are the ones to observe. In the cascode configuration, only the V_{GD} is the voltage to focus on, as it can show the highest voltage excursions. As mentioned above, the CS transistor experiences small voltage swings and the PA can be easily designed so this transistor is stressed below the breakdown limit. On the

H. Solar Ruiz and R. Berenguer Pérez, *Linear CMOS RF Power Amplifiers*,
DOI: 10.1007/978-1-4614-8657-2_5,

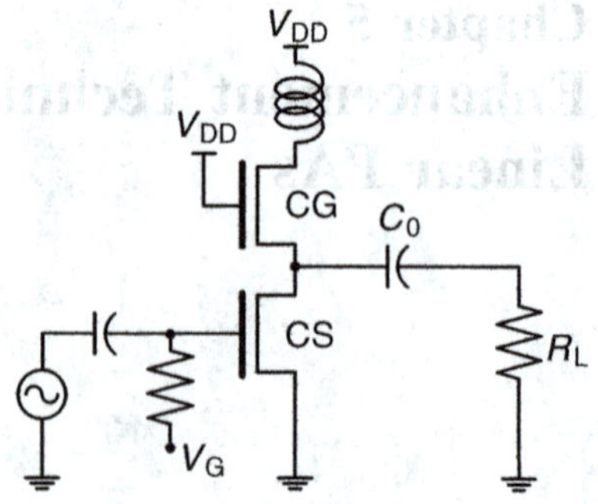

Fig. 5.1 Schematic view of a CMOS PA in the conventional cascode configuration

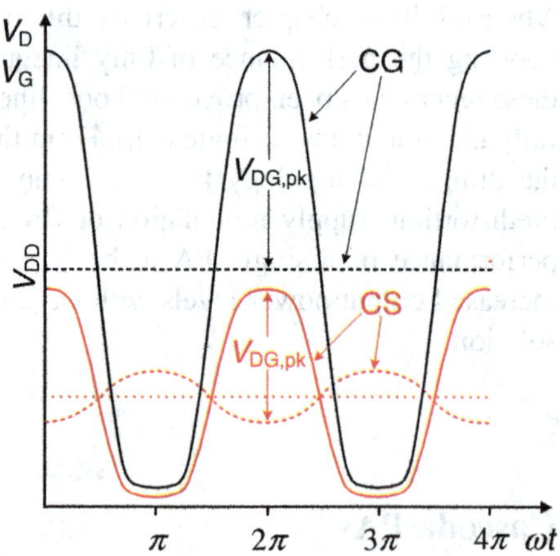

Fig. 5.2 V_G (*dashed line*) and V_D (*solid line*) waveforms of the CS (*red line*) and CG (*black line*) transistors of a conventional cascode PA

other hand, as the gate of the CG transistor is usually biased at V_{DD}, the cascode transistor is stressed on the gate-drain region by a voltage of approximately V_{DD} [1, 2], which is lower than a single transistor case. Therefore, as observed in Fig. 5.2, the gate oxide stress is shared by both the CS and the CG transistors so that the supply voltage can be then increased.

In terms of hot carrier degradation, the CG transistor sees the highest transistor stress, but for the same V_{DD} it is still lower than a single transistor case due to the presence of the CS transistor. More importantly, the peak voltage value occurs when the transistor current is at a minimum, which also helps to reduce the hot carrier effect [1].

The use of the CG transistor has another advantage; it improves the stability of the PA because the isolation between the input and the output of the stage increases [3]. This is particularly important when referring to power amplifiers, as the stability of the PA is crucial for a device dealing with large signals.

However, the main drawback of the cascode PA is the reduced headroom of the output voltage due to the fact that now two transistors are stacked between V_{DD}

and ground so that the overall V_{KNEE} is increased. This degrades the efficiency of the final PA if compared to a CS-only PA.

On the other hand, the voltage stress of the cascode PA is not equally shared by the CS and CG transistors. If V_{DG} is observed in Fig. 5.2, the V_{DG} voltage swing of the CG transistor becomes larger than that of the CS transistor. Therefore, the CG transistor is now the bottleneck in terms of PA reliability [2]. Still this imbalance of the cascode PA can be overcome by the following techniques.

Thick Oxide Cascode PAs

It is not unusual in standard CMOS processes to offer thick oxide transistor devices, usually with greater gate length so they are equivalent to the devices of the former CMOS processes. As discussed in Chap. 4, these thick oxide devices can tolerate much larger voltage swings. Therefore, a way to overcome the aforementioned limitation is to replace the CG transistor of the cascode PA with a thick oxide transistor so the reliability of the overall PA is ensured. Although this solution is used in practice [4–11], it has a drawback because the thick-oxide transistor presents worse high-frequency performance relative to the CMOS process standard device. At high frequencies, the thick-oxide device can then provide a lower gain at RF that should be compensated by, for example, increasing the gain of the driver stage [2, 4].

Self-Biased Cascode PAs

A different way to solve the imbalance issue of the conventional cascode PA is by properly biasing the gate of the CG transistor so that the stress level of both the CS and CG transistors is balanced [2, 5, 12–14]. This can be achieved by means of the modification in Fig. 5.3, which is referred to as the self-biased cascode configuration [2].

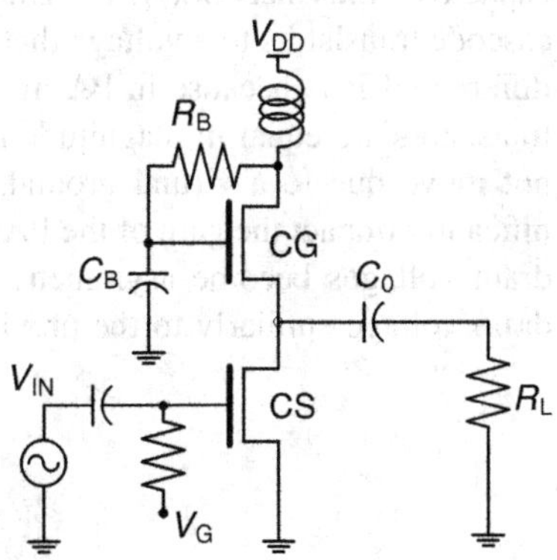

Fig. 5.3 Schematic view of a CMOS PA in the self-biased cascode configuration

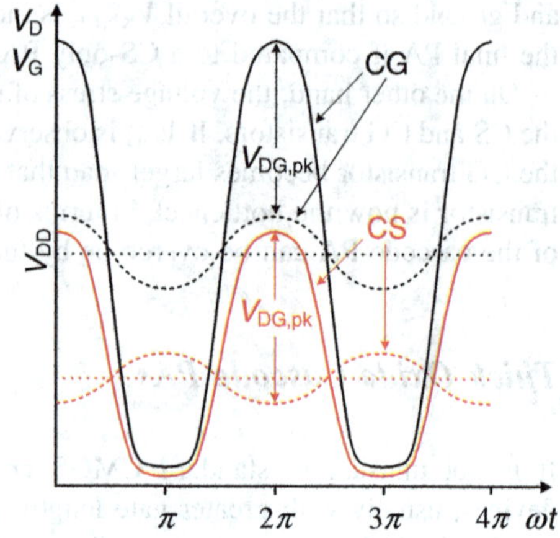

Fig. 5.4 V_G (*dashed line*) and V_D (*solid line*) waveforms of the CS (*red line*) and CG (*black line*) transistors of a self-biased cascode PA

Through the resistor R_B, the gate of the CG transistor tracks the output voltage. Capacitor C_B is added as a low pass filter in order to limit the voltage swing at the gate to a fraction of the output voltage. With such a configuration, the voltage swing at the gate of the CG transistor follows in phase with the output voltage. As a consequence, the V_{DS} of the CS transistor rises as well so that both transistors experience the same maximum V_{GD} voltage, as depicted in Fig. 5.4. However, the main drawback of this configuration is that as the gate voltage of the CG transistor follows the RF variation of the output voltage in both positive and negative swings around the gate DC value, a lower gain performance is obtained if compared to the conventional cascode configuration. Still the gain remains larger than that of the CG thick oxide transistor solution [2].

The self-biased cascode solution can be improved if a resistive-diode branch is connected in parallel to R_B so that the positive swings at the gate of the CG transistor can be made larger than the negative swings. Proper value of the resistor and size of the diode allow the average voltage of the CG transistor to be specified [2].

A different version of the previous resistive feedback network is to use just a capacitor between the drain and the gate of the cascode transistor acting as a capacitive bias network [11]. This gives the flexibility to bias the gate of the cascode transistors to a voltage that is different from V_{DD}. This scheme is useful in differential PAs because in PA linear operation the drain voltages of the cascode transistors are equal in magnitude and opposite in phase so their gate voltages do not move due to a virtual ground. Hence, the feedback capacitor does not significantly impact the gain of the PA. However, as the drain swing increases and the drain voltages become asymmetric, the cascode gate voltage starts tracking the drain voltage similarly to the previous configuration.

Multiple Stacked Cascode PAs

In principle, the stacking process could be extended to more than one CG transistor [15–19]. This would have the advantage of overcoming the breakdown voltage limit of CMOS process and increasing the V_{DD} and hence the output power levels.

In such a case, high output power levels can be achieved without requiring high output impedance transformation ratios so that the output matching network would present lower losses. This could compensate for the extra knee voltage of the stacked configuration so that efficiency can be kept high. In addition, the input impedance of the circuit is high compared to a single, wider transistor designed to offer the same output power levels, so the requirements of the driver stage are relaxed [20].

An example of a circuit using three stacked cascode transistors is shown in Fig. 5.5 [15, 16]. The resistive voltage divider ($R1$ to $R4$) is used to provide proper gate biasing of stacked transistors. In addition, this solution makes use of a capacitive voltage divider, which consists of the gate-to-source capacitance of each transistor and an external gate capacitor. The external capacitors are sized to enable the same drain-to-source, gate-to-source and drain-to-gate voltage swings for each transistor. These capacitors also adjust the input impedance of the middle and top transistors, so that the bottom and middle transistors see the optimum load impedance. The output voltage swings of all three transistors are combined in phase so that the voltage swing at the drain of top transistor is approximately three times higher than the drain-to-source voltage swing of each of the three transistors while the stress of transistors is still kept low.

However, this configuration has several limits. First, at high frequencies the gain is degraded due to presence of C_{gs} in the output impedance [15]. The three transistors will also see slightly different drain currents due to currents leaking out through C_{gs}. This makes the shaping of the voltage waveforms more difficult in

Fig. 5.5 Triple-stacked cascode PA

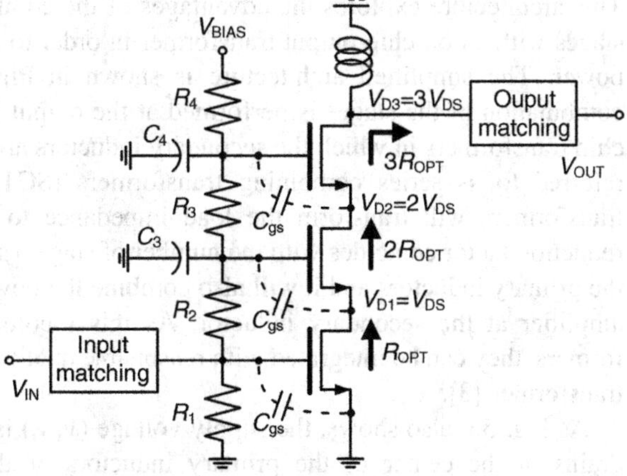

order to achieve high efficiency operation for all transistors simultaneously [15]. Secondly, in a standard CMOS process the bulk is connected to a fixed electrical potential (usually ground) and the drain-bulk diode experiences a reverse bias proportional to the absolute drain voltage. This diode has a limited reverse breakdown voltage. Although this breakdown voltage is relatively high in today's CMOS processes [21], the reverse breakdown voltage of the top transistor can become the limiting factor in the reliability of this PA type [15]. Finally, the performance of the PA is compromised by the source-body capacitance and body effect that progressively reduce the gain of the transistors in the upper sections of the stack [15].

Therefore, although examples of stacked cascode PAs can be also found in standard CMOS processes [18], this configuration is more suitable for CMOS process in which the bulk of the transistor is isolated as silicon-on-insulator (SOI), silicon-on-sapphire (SOS) or triple-well CMOS technologies [15–17].

Combined PAs

A different way to overcome the limits that a low V_{DD} imposes on the output power levels and the efficiency of a CMOS PA or to improve linearity is to combine several PAs in parallel. However, if a fully integrated PA is still desired, the power combination must be done carefully because the required output combining topology must show low losses in order to avoid degrading the efficiency of the overall PA. Several techniques allow PA combining while preserving the efficiency, the linearity and the output power of the resulting PA.

Integrated Transformer Power Combining

This architecture explores the advantages of the combination of several amplifier stages with an on-chip output transformer in order to increase the resulting output power. The simplified architecture is shown in Fig. 5.6. As can be seen, the combination of the stages is performed at the output by means of several 1:1 on-chip transformers in which the secondary inductors are connected in series. This is referred to as series combining transformers (SCT). The combination of the transformers will transform the load impedance to a lower value (ideally the reduction factor coincides with the number of stages) seen by each unit amplifier at the primary inductors and it will also combine the power generated from each unit amplifier at the secondary inductor. As this topology only requires 1:1 transformers, they can be integrated with reasonable quality factors if compared to a 1:n transformer [3].

As Fig. 5.6 also shows, the supply voltage (V_{DD}) is distributed to the transistor drains in the center of the primary inductors of the transformers and tuning

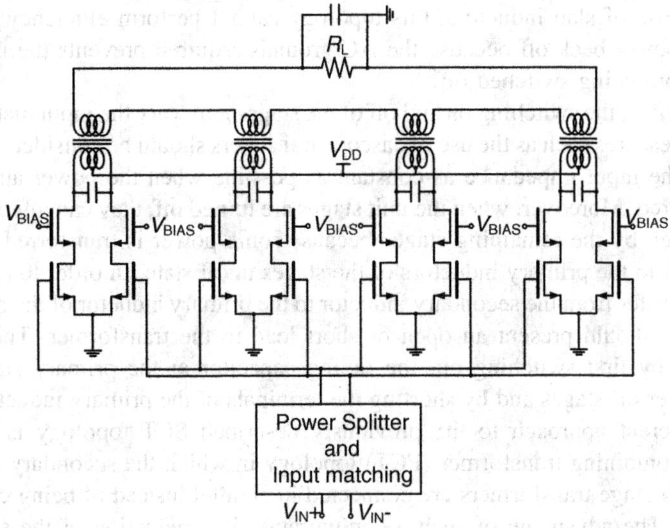

Fig. 5.6 Power combined PA

capacitors are placed in parallel to the transformer. The input network serves both to match the input driver and to split the input power to the different output stages. It can also be implemented as an input transformer [11, 22, 23] or LC matching network [24].

Although 1:1 transformers can be acceptable when integrated in efficient PAs, these transformers still show losses that must be minimized due to the high current flowing through them. A practical implementation tries to overcome the losses of the output transformers by making use of slab inductors [21, 22, 24], i.e. inductors that are formed by a straight piece of metal and are known as distributed active transformer (DAT). This solution allows inductors with high quality factors but at the cost of a low coupling factor between the primary and secondary inductors of the transformer [3]. Other solutions reduce the losses by using one-turn inductors [25, 26]. The layout of the output transformer can be modified to an 8-shape configuration and the use of double winding in the primary inductor for further loss reduction [26, 27]. However, if this last configuration is used, it must be noted that the primary inductors are not symmetric with respect to the secondary inductors and this introduces some amplitude and phase mismatch that must be taken into account [11].

In addition, the proposed transformer topology can dynamically modulate the load seen by each unit stage and so it is possible to keep the efficiency high at lower power levels. When the input power level is reduced to a certain power level, it is possible to switch off one or several of the output stages. In such a case the load seen by each of the on stages increases as the number of the on stages is reduced. The consequence is that output voltage returns to its maximum value and hence the efficiency of the overall PA is recovered [28]. It must be also noted that with the use of slab inductors, this topology cannot perform efficiency enhancement at power back-off because the AC grounds required prevents the individual stages from being switched off.

In practice, the switching on and off of the stages may vary the input matching, so certain measures such as the use of cascode transistors should be considered in order to keep the input impedance as constant as possible when the power amplifier is reconfigured. Moreover, when the unit stages are turned off, they can affect the load that is seen by the remaining stages because some power is transferred from the secondary to the primary inductors of the stages in off state. In order to reduce the power transfer from the secondary inductor to the primary inductor of the off stages, the stages should present an open or short load to the transformer. This can be achieved by first switching out the tuning capacitor at the primary side of the transformer off stages and by shorting the terminals of the primary inductor [28].

A different approach to the previously described SCT topology is to use a parallel combining transformer (PCT) topology in which the secondary inductors of the unit stage transformers are connected in parallel instead of being connected in series. The advantage of such a configuration is a reduction of the secondary inductor losses due to a reduction of the current flowing through them but at the cost of a larger area and a lower self-resonant frequency [29, 30]. Moreover, the PCT requires higher turn ratios in order to get the same output impedance compared to a SCT, which increases the losses in practice [11]. On the other hand, a combination of SCT and PCT topologies is also possible in order to combine the advantages of both transformer topologies in a parallel-series combining transformer (PSCT) topology [24]. The three topologies can be observed in Fig. 5.7.

The stability of this PA topology must be checked because a combination of internal feedback (due to transistor parasitics and drain inductance) and external feedback mechanisms (due to bypass capacitors, ground and supply line inductance) can lead to common mode oscillations. Potential instabilities can be solved by adding gate resistance, which degrades the gain, or by degrading the Q of the bypass capacitors [31].

Simple Parallel Combination

This topology combines two stages in the current domain with the aim of enhancing linearity and efficiency. Thus, as observed in Fig. 5.8, by simply connecting the output of a class A stage and class B stage the linearity and the efficiency of the combination can be enhanced compared to a single class A stage. This configuration takes advantage of the transconductance performance of the class B PAs [5, 6, 10]. One property of the class B stage is that its transconductance and resulting power gain increases with increasing input levels, just the opposite of the transconductance performance of a class A stage. Therefore, it is possible to combine both transconductances in order to improve the linearity of the resulting stage [6], as can be observed in Fig. 5.9.

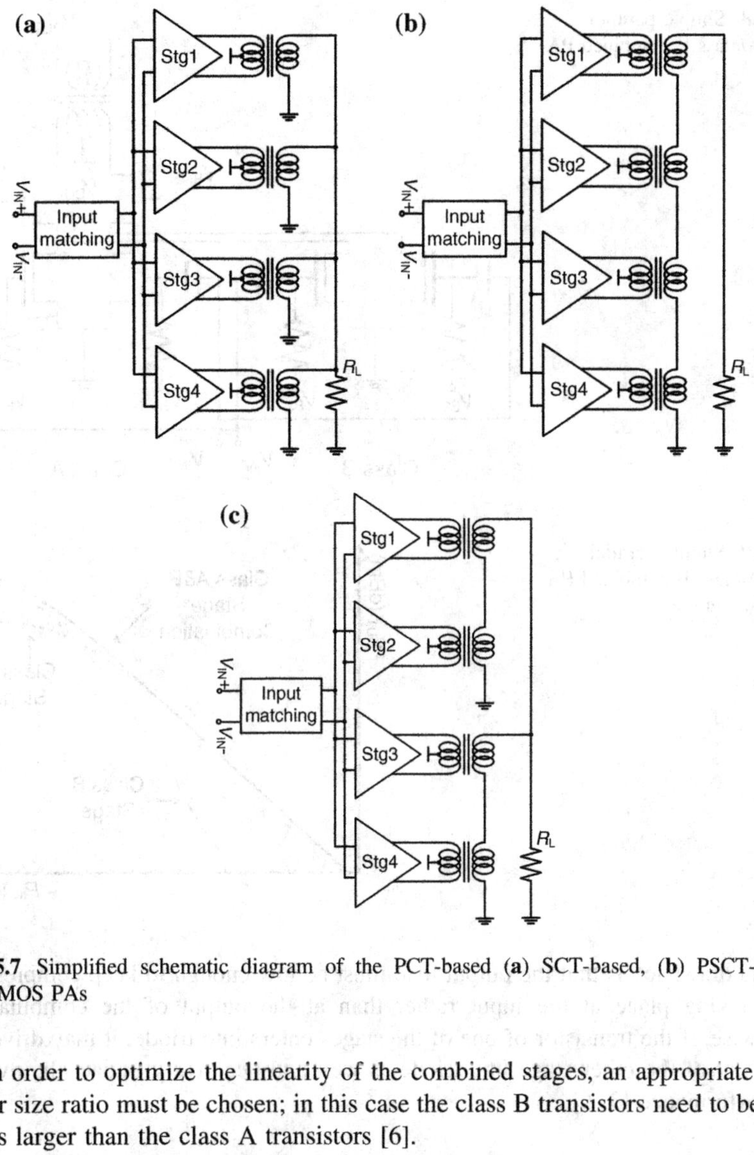

Fig. 5.7 Simplified schematic diagram of the PCT-based (**a**) SCT-based, (**b**) PSCT-based, (**c**) CMOS PAs

In order to optimize the linearity of the combined stages, an appropriate transistor size ratio must be chosen; in this case the class B transistors need to be four times larger than the class A transistors [6].

The efficiency of this configuration is almost the same as that of a class A amplifier at low power levels, and it closer to the efficiency of a class B amplifier at high power levels. This is because the two stages contribute to the majority of the gain and the current consumption at different input power ranges. At low power levels, the class B stage has almost no current consumption, whereas at high power levels the class B power contribution enhances efficiency.

This topology has several drawbacks, however. The first is that the gain of the parallel combination is low due to a lower overall effective transconductance. The

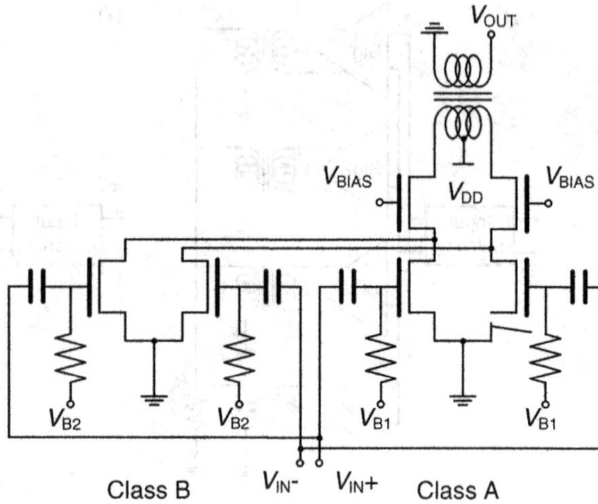

Fig. 5.8 Simple parallel class A/class B combined PA

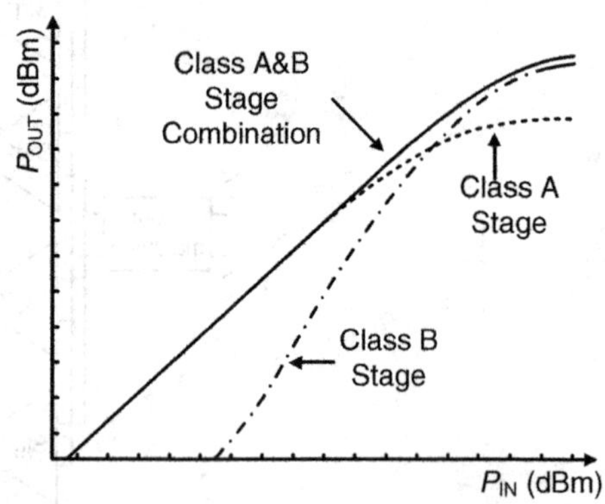

Fig. 5.9 Simple parallel class A/class B combined PA P_{IN}/P_{OUT} curve

second drawback is that the output load must be low enough to keep compression from taking place at the input rather than at the output of the combination; otherwise, if the transistor of one of the stages enters into triode, it may drive the transistor of the other stage into triode as well and negatively impact the overall PA performance [3, 6].

Doherty PAs

The Doherty power amplifier is an old concept. The first implementation was reported in 1936 [32], but recently it has attracted the attention due to its possibilities of being integrated. The Doherty power amplifier is an efficiency enhancement technique [33] and its simplified configuration is shown in Fig. 5.10.

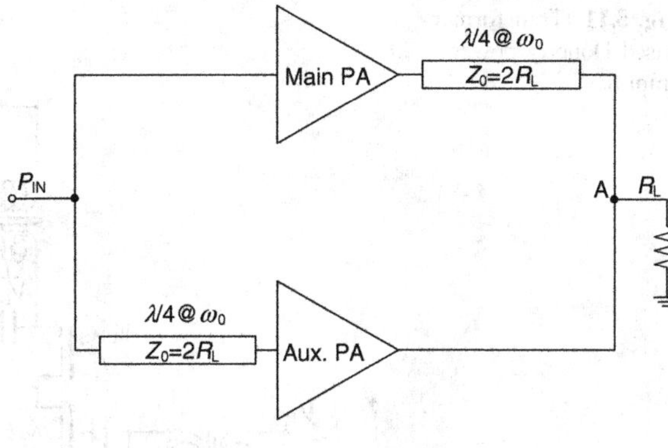

Fig. 5.10 Block diagram of the Doherty power amplifier

This architecture is based on the load modulation effect that the auxiliary amplifier (biased in class C) has on the main amplifier (biased in class A or AB). For low input drives, only the main amplifier contributes to the output power while seeing $4R_L$ due to the inverting effect of the output $\lambda/4$ transmission line. However when the input drive increases (usually to a power level in which the output power is 6 dB below P_{SAT}), the auxiliary amplifier turns on. The contribution of the auxiliary PA output current to the load causes the impedance seen at node A to increase. However, the output $\lambda/4$ transmission line acts as an impedance inverter, such that the main amplifier output impedance decreases up to $2R_L$. By controlling the amount of the auxiliary PA output current flowing through the load, it is possible to keep the output voltage swing of the main PA constant at its maximum level. Consequently, the main PA works at its maximum efficiency from the 6 dB back-off to P_{SAT} and the overall efficiency is kept high throughout the entire range. Finally, the input $\lambda/4$ transmission line serves to compensate the phase shift introduced by the output $\lambda/4$ transmission line.

In addition, the combination of the output powers of class AB and class C PAs ideally maintains the overall linearity of the resulting PA because the auxiliary amplifier shows gain extension behavior while the main amplifier enters into compression. At the peak power level, both amplifiers deliver the same amount of power as they both see a load of $2R_L$. A linearized P_{IN}-P_{OUT} curve is obtained until the auxiliary amplifier is saturated. It should be noted, however, that the conventional Doherty operation does not bring any enhancement in terms of increased output power levels since the maximum output power of both the class AB and class C amplifiers is the same as a single linear amplifier seeing the same output load R_L [23].

In order to keep efficiency high, the output matching circuit is a critical issue because the output network may introduce considerable loss that spoils the advantages of the Doherty PAs. In practice, only mm-wave Doherty PA designs implement the $\lambda/4$ transmission lines on-chip [34, 35]. At RF frequencies Doherty PAs cannot afford the large area of a $\lambda/4$ transmission line and therefore lumped components are used instead. Practical implementations of CMOS Doherty PAs at RF frequencies make use of an integrated LC π network in order to emulate the

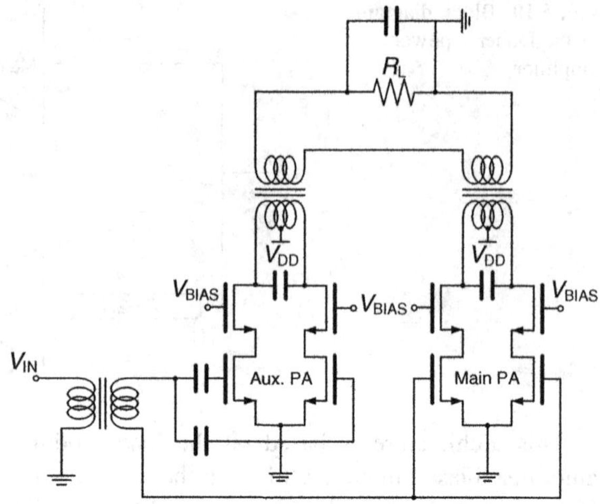

Fig. 5.11 Transformer-based Doherty power amplifier

behavior of the output $\lambda/4$ transmission line while saving area [13, 14, 36–38]. However, most of them make use of off-chip components or the bondwire for the LC π network implementation [36–38] in order to preserve the efficiency of the PA because the efficiency of CMOS Doherty PAs with integrated lumped components is low [14]. Some enhancement can be achieved if slab inductors are used [13] to compose the LC π network, but at the cost of larger area and modest efficiency enhancement [23]. It must be also noted that an extra network is still necessary along with the $\lambda/4$ transmission line in order to downconvert the 50 Ω output load to the optimum load of the Doherty PA (the value of R_L is chosen to be half of the R_{OPT}).

As shown in Fig. 5.11, recent Doherty PA designs substitute the $\lambda/4$ output inverter at the output with SCT transformers [23, 39, 40]. As the SCT solution provides the required load modulation that is the basis of the Doherty amplifier, the output $\lambda/4$ transmission line can be eliminated while preserving the crucial characteristic of Doherty PAs, i.e. efficiency enhancement at back-off. An enhanced SCT Doherty solution that further improves the efficiency at back-off is to use asymmetric SCT transformers at the output. This solution improves the efficiency at power back-off compared to previous Doherty PAs because it alleviates the impedance matching loss at back-off for the conventional integrated Doherty architecture or the symmetrical SCT solution [23].

Finally, the size ratio, transistor matching and biasing of both main and auxiliary PAs must be carefully chosen because linearity is an issue in the region where the auxiliary PA start to operate [23, 41]. If the combination of these two PAs in the Doherty power amplifier is not properly carried out, the PA will not be able to fulfill the stringent ACLR specifications of modern communication standards even at low power levels.

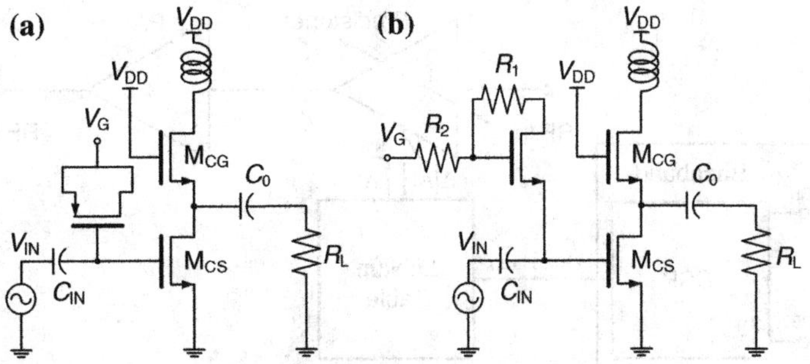

Fig. 5.12 Analog RF predistortion based on **(a)** A PMOS transistor and **(b)** NMOS diode with resistor feedback

Predistorted PAs

Predistortion is a linearization technique in which the input RF signal is distorted in such a way that this distortion cancels the distortion of the signal due to the nonlinearities of the amplification process. In linear RF CMOS PAs this technique is applied by means of several different solutions in order to correct the AM-AM and the AM-PM distortions introduced by the amplifier.

A first approach is analog predistortion, where the aim is to correct mainly the AM-PM distortion of the NMOS amplifying transistor by means of a circuit connected in parallel to the transistor bias. This circuit ideally cancels the phase distortion that the nonlinear NMOS transistor gate-to-source capacitor introduces at increasing input power levels by adding a circuit showing an opposite capacitance variation with input power. Some practical implementations use a PMOS transistor connected in parallel to the gate of the NMOS transistor [11, 42–44] or an NMOS transistor with feedback resistors [45–47], as can be observed in Fig. 5.12.

Other solutions that can be implemented are NMOS transistors acting as varactors [48], a series linearizer based on NMOS transistors acting as variable resistors [25] or feedback techniques [49]. In general, although analog predistortion introduces little extra complexity to the PA, its enhancement in terms of linearity is poor and subject to process dispersions or changing PA conditions, such as temperature or load impedance variations.

As observed in Fig. 5.13, more complex solutions are baseband digital predistorters [9] or digitally assisted RF predistorters [50, 51], in which the amplitude of the input signal feeds a lookup table (LUT) that controls the predistortion circuit. These techniques show better performance than the previous ones while achieving highly integrated implementations. However, because they are still open-loop solutions, they are not able to adapt to changing PA conditions and PA memory effects [33]. In addition to this, extra memory and some digital signal processing (DSP) is required in the baseband circuitry.

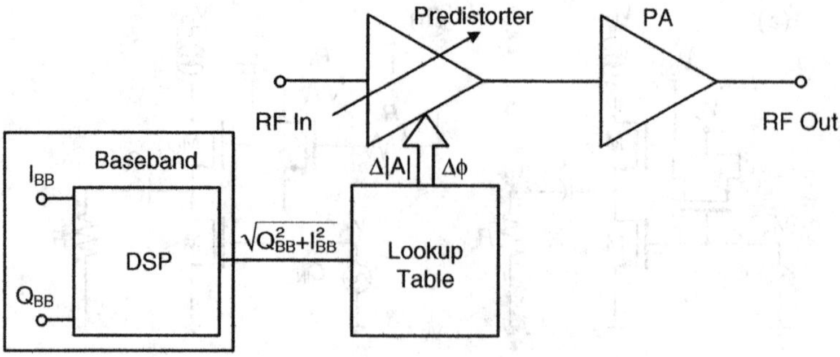

Fig. 5.13 Digitally assisted RF predistorted PA architecture

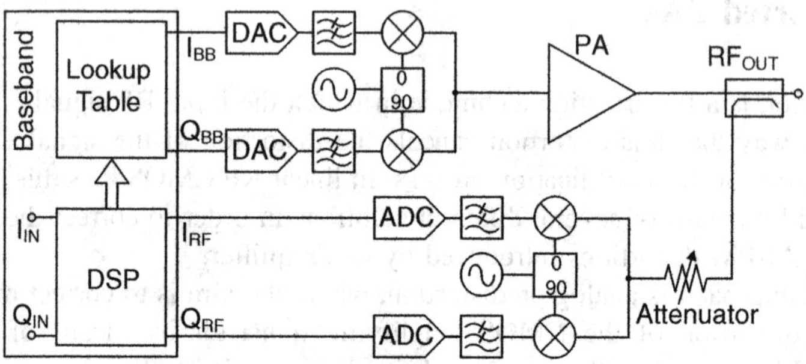

Fig. 5.14 Adaptive digital baseband predistorted PA architecture

A further step in enhancing the linearity of linear PAs based on predistortion is to use closed-loop techniques so that the predistortion circuit may adapt to the changing conditions of PAs, as shown in Fig. 5.14 [10, 52, 53]. In these techniques the LUT is fed by an attenuated version of the RF output signal that is down-converted to baseband. Therefore, the predistortion circuit is controlled by the actual signal at the PA output. This last solution offers much better linearity enhancement, but at the cost of increased complexity since an extra receiver for the feedback path along with more powerful DSP capabilities and memory for the LUT at the baseband are required.

Dynamic Supply or Envelope Tracking

Dynamic supply or envelope tracking is an efficiency enhancement technique based on the older envelope elimination and restoration (EER) architecture that was proposed by Kahn in 1952 [54]. As Fig. 5.15 shows, the EER architecture has a secondary path on which the envelope of the input RF signal is first detected and

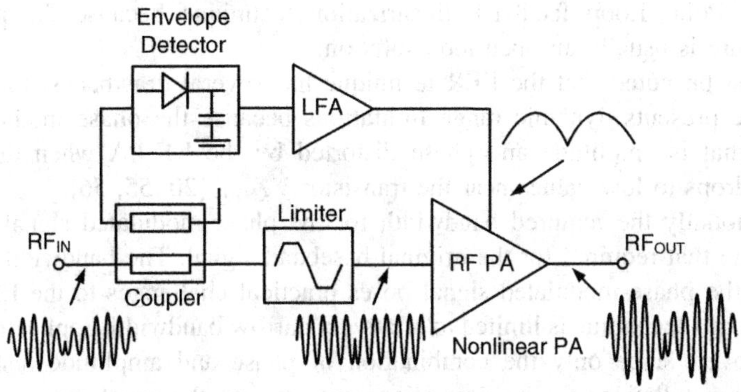

Fig. 5.15 Envelope elimination and restoration (EER) architecture

further amplified by a highly efficient low-frequency amplifier (LFA), which is also referred to as a modulator.

On the main path the RF input signal passes through a limiter that removes the envelope but preserves the phase modulation of the signal. The phase-modulated signal can then be amplified by a highly efficient nonlinear RF PA because the limiter eliminates the possibility of AM-PM distortion in the PA. Finally, the envelope signal and the amplified phase modulated RF signal are combined at the PA by modulating the supply voltage so that amplitude modulation is restored. The main advantage of the EER technique then is the possibility of using highly efficient nonlinear switched-mode RF PAs for variable envelope modulation channels instead of low-efficiency linear RF PAs.

In modern fully integrated transmitters, there is no need to first create the RF signal so it can be split up into a phase and envelope signal. Instead, the amplitude and phase information is provided separately in the baseband by a DSP block. This modern version of the classic EER, shown in Fig. 5.16, is also known as Polar PA because the baseband signal is converted to a polar format, i.e. amplitude and phase separation, before being amplified. However, this should not be confused

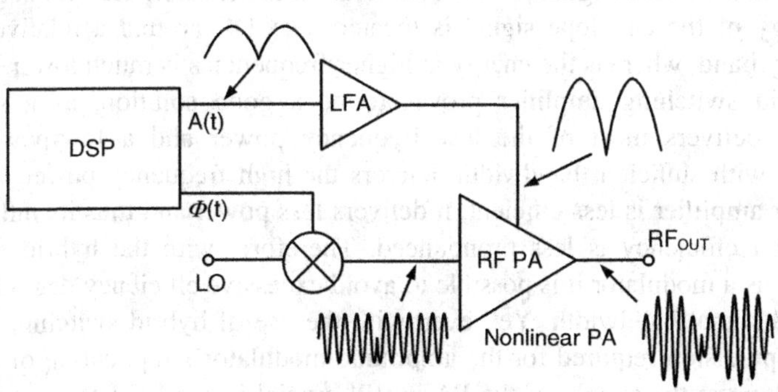

Fig. 5.16 Polar PA architecture

with the Polar Loop feedback linearization technique because the polar PA architecture is usually an open-loop solution.

It must be noted that the EER technique has several drawbacks. Firstly, this technique presents dynamic range limitations because the phase modulated RF input signal is amplitude and phase distorted by the RF PA when the supply voltage drops to low values near the transistor V_{KNEE} [20, 55, 56].

Additionally the required bandwidth for the phase-modulated signal is much wider than that required for the original baseband signal. The bandwidth requirement of the phase-modulated signal poses practical challenges to the EER technique, so this technique is limited to relatively narrow bandwidth applications [57].

Moreover, since only the combination of phase and amplitude restores the original constellation, precise time alignment between the envelope and RF paths is required [57, 58]. For the same reason, the LFA or modulator must generate a suitable supply voltage with great accuracy, requiring both high linearity and great bandwidth. Along with this, the total system efficiency is determined by the product of the envelope amplifier efficiency and the RF PA efficiency, which means that a high-efficiency modulator is critical to the EER technique. High-efficiency modulators can be realized in practice with DC/DC converters, but the switching frequency needs to be several times the signal bandwidth [57]. The reason is that the stringent requirements of the EER technique allow only a very low ripple at their output. A low-ripple DC/DC converter requires a switching frequency that is much higher than the maximum envelope bandwidth so the low-pass filter can provide a high attenuation of the ripple while passing the signals within the envelope bandwidth. Traditional high-efficiency switching-mode DC/DC converters are realized based on delta modulation or pulse width modulation (PWM) techniques, where a high switching frequency will introduce a significant switching loss. In such cases, the efficiency of these modulators, and hence the efficiency of the whole architecture, is greatly degraded when the signal bandwidth is increased.

In order to overcome the issue of efficiency degradation of the modulator for wideband signals, the hybrid switching amplifier is implemented instead [58]. This type of amplifier consists of a combination of a wideband linear amplifier and a high-efficiency switching amplifier. The solution is based on the fact that most of the energy of the envelope signal is located near DC around a relatively small frequency band, whereas the energy at higher frequencies is much lower. As such, the hybrid switching amplifier proves to be a good solution, as a switching amplifier delivers most of the low frequency power and a low-power linear amplifier with sufficient bandwidth delivers the high frequency power. Although the linear amplifier is less efficient, it delivers less power and thus its influence on the overall efficiency is less pronounced. Therefore, with the hybrid switching amplifier as a modulator it is possible to avoid excessive efficiency degradation for increased signal bandwidth. Yet, even with the use of hybrid switching modulators, the precision required for the amplitude modulator's replication of the input RF envelope at the supply of the PA is still crucial in the EER technique.

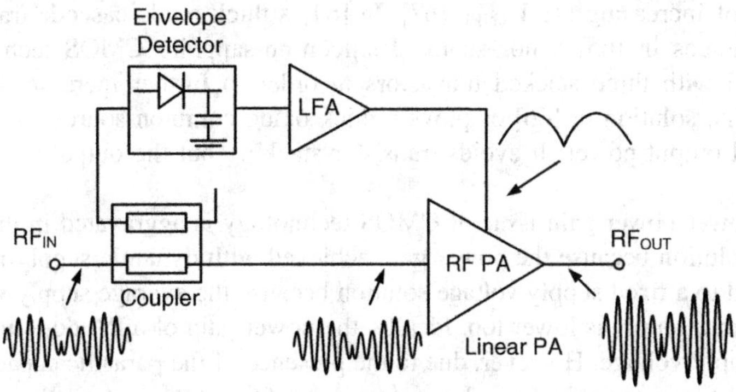

Fig. 5.17 Envelope tracking PA architecture

One way to relax the requirements imposed on the modulator by the EER technique is the use of another efficiency enhancement technique called dynamic supply or envelope tracking (ET), illustrated in Fig. 5.17. This technique relaxes the EER technique issues by removing the limiter from the RF path so that the phase and amplitude modulation are kept in the input PA signal. As the input signal preserves the amplitude and phase information, in this technique the envelope signal is not used to modulate the RF PA but rather it is only to provide a DC supply level that is just enough for the RF PA to amplify the input RF signal without compression [59]. This, of course, requires the use of a less efficient linear PA, but this can be compensated by the fact that the ET requires a lower modulator bandwidth and less precise time alignment between the envelope and RF paths [57]. In addition, the RF signal bandwidth in the ET architecture is identical to the baseband signal bandwidth, which is much narrower than the phase modulated RF signal bandwidth required in the EER system.

On the other hand, the dynamic range issues of the EER technique can be essentially removed in the ET technique by allowing for normal linear operation at a reduced supply voltage, i.e. avoiding a voltage supply below the V_{KNEE}, in the small signal regime [33, 56].

Practical implementations of the ET architecture in CMOS processes achieves high efficiency hybrid modulators, though they are usually applied to GaAs or SiGe PAs [56, 58, 60–63] and only a limited number of implementations look for the complete PA integration in a unique CMOS technology [64–66]. The reason may come from the fact that the CMOS technology presents two major issues in this case: first, the lower breakdown voltage due to the scaling down in technology that restricts the maximum V_{DD} along with relatively high V_{KNEE}; and second, the reduced transconductance of MOS transistors leading to inferior power gain and poorer efficiency [66]. Both problems are crucial in ET PAs. Low V_{DD} and high V_{KNEE} prevent a CMOS ET PA from having a high dynamic range because the V_{DD} to V_{KNEE} voltage range is reduced. A higher dynamic range can be achieved by stacking a cascode transistor, as is the case in [64] and [65], but this comes with the cost of increasing the V_{KNEE} [67]. In [64] a thick oxide cascode transistor is used, whereas in [65] a non-standard silicon-on-sapphire CMOS technology is employed with three stacked transistors in order to further increase the output power. The solution in [66] employs a thick oxide common-source transistor for increased output power. It avoids transistor stacking but the output power levels are low.

The lower power gain issue of CMOS technology is aggravated in the case of the ET solution because the power gain achieved with dynamic supply is inferior compared to a fixed supply voltage solution because the average supply voltage in the dynamic supply is lower too. Ideally, the power gain of a PA does not depend on its supply voltage. However, due to the presence of the parasitic capacitance of the power transistor, this dependence does exist [66]. Moreover, as the crest factor of the signal increases, the average supply voltage applied to the PA decreases. This induces an even lower average gain, and the consequence is that the overall PAE is reduced [58]. In fact, the solution implemented in standard CMOS technologies shows low power gains [64, 66].

Finally, it must be noted that the ET PA architecture can be also combined with some of the previously discussed solutions, for example, the ET Doherty PA architecture [58].

Adaptive Biasing

The adaptive biasing architecture is both an efficiency and linearity enhancement technique in which the transistor bias is adapted to the amplitude of the input RF signal so that the RF PA operates in different classes according to envelope amplitude. Fig. 5.18 shows a simple adaptive biasing scheme.

Efficiency enhancement comes from the fact that in modern digital communication standards, the signal works at its peak power level during a small percentage of its total operation period [68]. Therefore, the RF PA wastes a significant amount

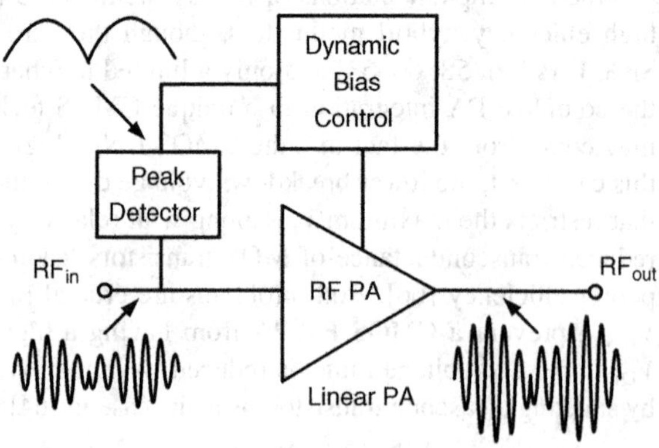

Fig. 5.18 Bias adaptation PA architecture

of power if the biasing current is kept high, independently of the signal amplitude. In the adaptive biasing technique the RF PA is biased at a high quiescent point to provide linear operation only in the high power region. However, it is not power efficient to maintain the high bias condition when the transmitted power is low. In the low power case, it is possible to ensure the linearity performance of the PA while biasing the PA at lower bias conditions. Consequently, as the main increase of bias occurs near P_{1dB}, the adaptive biasing reduces the average current from supply and hence enhances PA efficiency, as shown in Fig. 5.19.

Linearity is somewhat improved in this technique [67]. The P_{1dB} of a PA with dynamic biasing is often higher than that obtained at fixed bias, as shown in Fig. 5.19. This is attributed to the nonlinear nature of CMOS transistor transconductance where transconductance increases with increases in the gate bias, and therefore the PA with dynamic biasing has a higher linear range. However, as transconductance is low in the low power region, the linearity enhancement brings with it the issue that the resulting power gain is lower than in the fixed bias scheme. It must be also noted that as the adaptive biasing scheme of Fig. 5.18

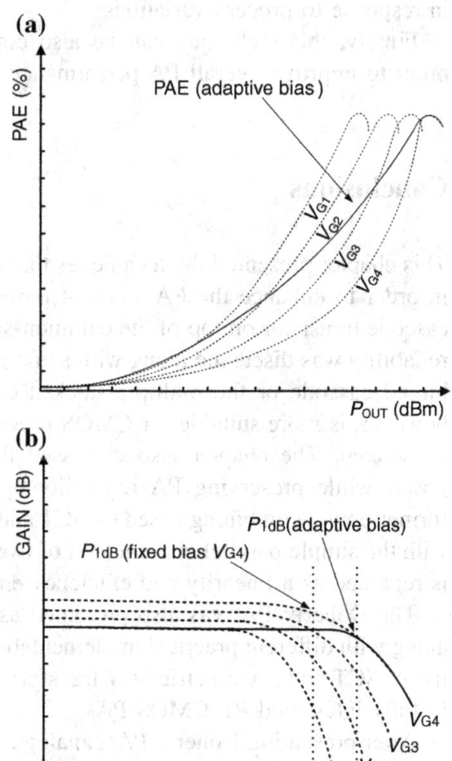

Fig. 5.19 **a** PAE and **b** P_{1dB} improvement of a CMOS RF PA when an adaptive biasing is applied

senses the envelope of the input signal, the linearity enhancement is degraded at the PA changing conditions due to gain, temperature or process dispersions.

Similar to the envelope tracking architecture, in adaptive biasing the bandwidth of the envelope tracking circuit should be high enough to track the input channel bandwidth. However, this technique has faster tracking speed to accommodate the wideband input signals [67].

Regarding stability, the PA may need an additional stabilizing resistor, such as shunt feedback between gate and drain. In fact, implemented adaptive biased PAs with a cascode transistor usually require this stabilizing solution [69, 70].

This architecture is been proven useful especially for CMOS PAs working at frequencies of tens of GHz [68, 70–72]. In such high frequency applications, complex solutions are not so useful. For example, the improvement in efficiency is not significant when using the Doherty structure compared to the conventional cascode structure [71]. Moreover, the previously presented simple parallel combination is not very efficient in the mm-Wave regime because class B bias lowers the device speed. Again, analog predistortion based on the capacitance compensation technique induces extra capacitance load and hinders its employment in high frequency regimes [68]. It must be noted than in [72], the peak detector senses the output signal so that the feedback loop may also adjust the output signal in response to process variations.

Finally, this technique can be also combined with previous architectures in order to improve overall PA performance [36, 73].

Conclusions

This chapter presented the techniques that can be applied to linear CMOS RF PAs in order to enhance the PA's output power, linearity and efficiency. Placing the cascode transistor on top of the common-source transistor in order to improve PA reliability was discussed along with some possible enhancements, such as the self-biased cascode or the multiple stacked cascode transistors. This last technique, however, is more suitable for CMOS processes in which the bulk of the transistor is isolated. The chapter also discussed the possibility of increasing the output power while preserving PA reliability by combining several stages. The transformer power combining based on SCT and PCT transformers was presented along with the simple parallel combination of two different PA classes. This last solution is regarded as a linearity and efficiency enhancement technique.

The Doherty PA was also presented as an efficiency enhancement technique, along with different practical implementations of the $\lambda/4$ transmission lines and the use of SCT and asymmetric SCT transformers in order to achieve high efficiencies in fully integrated RF CMOS PAs.

After presenting Doherty PAs, analog and digital predistortion techniques were discussed along with the trade-off between linearity enhancement and complexity. The envelope tracking technique was also addressed and confronted with the classical EER technique. The ET technique has looser modulator and bandwidth requirements and thus is a better compromise for the implementation of efficiency enhanced CMOS PAs. The specific issues regarding the CMOS technology in the implementation of a complete CMOS ET PA were also covered. Finally, using the adaptive biasing technique as an efficiency and linearity enhancement technique was discussed along with the value of using this technique in CMOS PAs working at mm-Wave frequencies.

References

1. Niknejad AM, Chowdhury D, Jiashu C (2012) Design of CMOS power amplifiers. IEEE Trans Microw Theory Techn 60(6):1784–1796
2. Sowlati T, Leenaerts DMW (2003) A 2.4 GHz 0.18 μm CMOS self-biased cascode power amplifier. IEEE J Solid-State Circuits 38(8):1318–1324
3. Razavi B (2011) RF microelectronics, 2nd edn. Prentice Hall, New Jersey
4. Hamhee J, Kun-Seok L, Ockgoo L, Kyu Hwan A, Youngchang Y, Hyungwook K, Dong Ho L, Jongsoo L, Chang-Ho L, Laskar J (2010) A 40 % PAE linear CMOS power amplifier with feedback bias technique for WCDMA applications. In: IEEE radio frequency integrated circuits symposium (RFIC 2010), pp 561–564
5. Jian F, Shilei H, Yumei H, Zhiliang H (2010) A 2.4 GHz CMOS power amplifier. In: IEEE International Conference Solid-State and Integrated Circuit Technology (ICSICT 2010), pp 659–661
6. Yongwang D, Harjani R (2005) A high-efficiency CMOS +22 dBm linear power amplifier. IEEE J Solid-State Circuits 40(9):1895–1900
7. Francois B, Reynaert P (2012) A fully integrated watt-level linear 900 MHz CMOS RF power amplifier for LTE-applications. IEEE Trans Microw Theory Techn 60(6):1878–1885
8. Nuyts PAJ, Francois B, Dehaene W, Reynaert P (2012) A CMOS burst-mode transmitter with watt-level RF PA and flexible fully digital front-end. IEEE Trans Circuits Syst II: Exp Briefs 59(10):613–617
9. Degani O, Cossoy F, Shahaf S, Cohen E, Kravtsov V, Sendik O, Chowdhury D, Hull CD, Ravid S (2010) A 90 nm CMOS power amplifier for 802.16e (WiMAX) applications. IEEE Trans Microw Theory Techn 58(5):1431–1437
10. Afsahi A, Behzad A, Magoon V, Larson LE (2010) Linearized dual-band power amplifiers with integrated baluns in 65 nm CMOS for a 2×2 802.11n MIMO WLAN SoC. IEEE J Solid-State Circuits 45(5):955–966
11. Chowdhury D, Hull CD, Degani OB, Yanjie W, Niknejad AM (2009) A fully integrated dual-mode highly linear 2.4 GHz CMOS power amplifier for 4G WiMax applications. IEEE J Solid-State Circuits 44(12):3393–3402
12. Apostolidou M, Van der Heijden MP, Leenaerts DMW, Sonsky J, Heringa A, Volokhine I (2009) A 65 nm CMOS 30 dBm Class-E RF power amplifier with 60 % PAE and 40 % PAE at 16 dB back-off. IEEE J Solid-State Circuits 44(5):1372–1379
13. Elmala M, Paramesh J, Soumyanath K (2006) A 90 nm CMOS Doherty power amplifier with minimum AM-PM distortion. IEEE J Solid-State Circuits 41(6):1323–1332
14. Li-Yuan Y, Hsin-Shu C, Yi-Jan C (2008) A 2.4 GHz fully integrated cascode-cascade CMOS Doherty power amplifier. IEEE Microw Wireless Compon Lett 18(3):197–199
15. Pornpromlikit S, Jeong J, Presti CD, Scuderi A, Asbeck PM (2010) A watt-level stacked-FET linear power amplifier in silicon-on-insulator CMOS. IEEE Trans Microw Theory Techn 58(1):57–64

16. Jeong J, Pornpromlikit S, Asbeck PM, Kelly D (2006) A 20 dBm linear RF power amplifier using stacked silicon-on-sapphire MOSFETs. IEEE Microw Wireless Compon Lett 16(12):684–686
17. Leuschner S, Pinarello S, Hodel U, Mueller JE, Klar H (2010) A 31 dBm, high ruggedness power amplifier in 65 nm standard CMOS with high-efficiency stacked-cascode stages. In: IEEE radio frequency integrated circuits symposium (RFIC 2010), pp 395–398
18. Mingyuan L, Afsahi A, Behzad A (2010) A single-chip 2.4 GHz double cascode power amplifier with switched programmable feedback biasing under multiple supply voltages in 65 nm CMOS for WLAN application. In: IEEE radio frequency integrated circuits symp (RFIC 2010), pp 391–394
19. Ezzeddine AK, Huang HC (2003) The high voltage/high power FET (HiVP). In: IEEE radio frequency integrated circuits symppsium (RFIC 2003), pp 215–218
20. Reynaert P, Steyaert M (2006) RF power amplifiers for mobile communications. Springer, The Netherlands
21. Aoki I, Kee S, Magoon R, Aparicio R, Bohn F, Zachan J, Hatcher G, McClymont D, Hajimiri A (2008) A fully-integrated quad-band GSM/GPRS CMOS power amplifier. IEEE J Solid-State Circuits 43(12):2747–2758
22. Aoki I, Kee SD, Rutledge DB, Hajimiri A (2002) Fully integrated CMOS power amplifier design using the distributed active-transformer architecture. IEEE J Solid-State Circuits 37(3):371–383
23. Kaymaksut E, Reynaert P (2012) Transformer-based uneven Doherty power amplifier in 90 nm CMOS for WLAN applications. IEEE J Solid-State Circuits 47(7):1659–1671
24. Jihwan K, Woonyun K, Hamhee J, Yan-Yu H, Youngchang Y, Hyungwook K, Chang-Ho L, Kornegay KT (2012) A fully-integrated high-power linear CMOS power amplifier with a parallel-series combining transformer. IEEE J Solid-State Circuits 47(3):599–614
25. Onizuka K, Ishihara H, Hosoya M, Saigusa S, Watanabe O, Otaka S (2012) A 1.9 GHz CMOS power amplifier with embedded linearizer to compensate AM-PM distortion. IEEE J Solid-State Circuits 47(8):1820–1827
26. Haldi P, Chowdhury D, Reynaert P, Gang L, Niknejad AM (2008) A 5.8 GHz 1 V linear power amplifier using a novel on-chip transformer power combiner in standard 90 nm CMOS. IEEE J Solid-State Circuits 43(5):1054–1063
27. Boshi J, Moon J, Chenxi Z, Kim B (2012) A 30.8 dBm wideband CMOS power amplifier with minimized supply fluctuation. IEEE Trans Microw Theory Techn 60(6):1658–1666
28. Gang L, Haldi P, Tsu-Jae King L, Niknejad AM (2008) Fully integrated CMOS power amplifier with efficiency enhancement at power back-off. IEEE J Solid-State Circuits 43(3):600–609
29. Afsahi A, Larson LE (2010) An integrated 33.5 dBm linear 2.4 GHz power amplifier in 65 nm CMOS for WLAN applications. In: IEEE Custom Integrated Circuits Conference (CICC 2010), pp 1–4
30. Kyu Hwan A, Dong-Ho L, Ockgoo L, Hyungwook K, Jeonghu H, Woonyun K, Chang-Ho L, Haksun K, Laskar J (2009) A 2.4 GHz fully integrated linear cmos power amplifier with discrete power control. IEEE Microw Wireless Compon Lett 19(7):479–481
31. Chowdhury D, Reynaert P, Niknejad AM (2008) Transformer-coupled power amplifier stability and power back-off analysis. IEEE Trans Circuits Syst II: Exp Briefs 55(6):507–511
32. Doherty WH (1936) A new high efficiency power amplifier for modulated waves. Proc Inst Radio Engineers 24(9):1163–1182
33. Cripps SC (1999) RF power amplifiers for wireless communications. Artech House, Norwood
34. Agah A, Hanafi B, Dabag H, Asbeck P, Larson L, Buckwalter J (2012) A 45 GHz Doherty power amplifier with 23 % PAE and 18 dBm output power, in 45 nm SOI CMOS. In: IEEE MTT-S international microwave symposium digest (IMS 2012), pp 1–3
35. Wicks B, Skafidas E, Evans R (2008) A 60 GHz fully-integrated Doherty power amplifier based on 0.13 mm CMOS process. In: IEEE radio frequency integrated circuits symposium (RFIC 2008), pp 69–72

36. Chen YJE, Chih-Yun L, Tang-Nian L, Heo D (2006) A high-efficient CMOS RF power amplifier with automatic adaptive bias control. IEEE Microw Wireless Compon Lett 16(11):615–617
37. Wongkomet N, Tee L, Gray PR (2006) A +31.5 dBm CMOS RF Doherty power amplifier for wireless communications. IEEE J Solid-State Circuits 41(12):2852–2859
38. Kang J, Daekyu Y, Kyoungjoon M, Kim B (2006) A ultra-high PAE doherty amplifier based on 0.13 μm CMOS process. IEEE Microw Wireless Compon Lett 16(9):505–507
39. Kaymaksut E, Reynaert P (2010) A 2.4 GHz fully integrated Doherty power amplifier using series combining transformer. In: Proceedings of european solid-state circuits conference (ESSCIRC 2010), pp 302–305
40. Gaber WM, Wambacq P, Craninckx J, Ingels M (2012) A CMOS IQ digital doherty transmitter using modulated tuning capacitors. In: Proceedings of european solid-state circuits conference (ESSCIRC 2012), pp 341–344
41. Yu Z, Iwamoto M, Larson LE, Asbeck PM (2003) Doherty amplifier with DSP control to improve performance in CDMA operation. In: IEEE MTT-S international microwave symposium digest (IMS 2003), pp 687–690
42. Chengzhou W, Vaidyanathan M, Larson LE (2004) A capacitance-compensation technique for improved linearity in CMOS class-AB power amplifiers. IEEE J Solid-State Circuits 39(11):1927–1937
43. Chengzhou W, Larson LE, Asbeck PM (2001) A nonlinear capacitance cancellation technique and its application to a CMOS class AB power amplifier. In: IEEE radio frequency integrated circuits symposium (RFIC 2001), pp 39–42
44. Chao L, Pham AVH, Shaw M, Saint C (2007) Linearization of CMOS broadband power amplifiers through combined multigated transistors and capacitance compensation. IEEE Trans Microw Theory Techn 55(11):2320–2328
45. Xian C, Roblin P, Jongsoo L, Young-Gi K, Young Gi K (2007) A 3.5 GHz CMOS Doherty power amplifier with integrated diode linearizer targeted for WiMax applications. In: IEEE midwest symposium on circuits and systems (MWSCAS 2007), pp 465–468
46. Cheng-Chi Y, Huey-Ru C (2003) A 0.25-μm 20 dBm 2.4 GHz CMOS power amplifier with an integrated diode linearizer. IEEE Microw Wireless Compon Lett 13(2):45–47
47. Singh RD, Kyung-Wan Y (2006) A linear mode CMOS power amplifier with self-linearizing bias. In: IEEE Asian Solid-State Circuits Conference (ASSCC 2006), pp 251–254
48. Ding M, Gard KG, Steer MB (2012) A highly linear and efficient CMOS RF power amplifier with a 2-D circuit synthesis technique. IEEE Trans Microw Theory Techn 60(9):2851–2862
49. Hamhee J, Kun-Seok L, Ockgoo L, Kyu Hwan A, Youngchang Y, Hyungwook K, Kobayashi KW, Chang-Ho L, Kenney JS (2013) A cascode feedback bias technique for linear CMOS power amplifiers in a multistage cascode topology. IEEE Trans Microw Theory Techn 61(2):890–901
50. Palaskas Y, Taylor SS, Pellerano S, Rippke I, Bishop R, Ravi A, Lakdawala H, Soumyanath K (2006) A 5 GHz 20 dBm power amplifier with digitally assisted AM-PM correction in a 90 nm CMOS process. IEEE J Solid-State Circuits 41(8):1757–1763
51. Ki Yong S, Bonhoon K, Songcheol H (2012) A CMOS power amplifier with a built-in RF predistorter for handset applications. IEEE Trans Microw Theory Techn 60(8):2571–2580
52. Kwon DH, Hao L, Yuchun C, Tseng R, Yun C (2010) Digitally equalized cmos transmitter front-end with integrated power amplifier. IEEE J Solid-State Circuits 45(8):1602–1614
53. Presti CD, Carrara F, Scuderi A, Asbeck PM, Palmisano G (2009) A 25 dBm digitally modulated CMOS power amplifier for WCDMA/EDGE/OFDM with adaptive digital predistortion and efficient power control. IEEE J Solid-State Circuits 44(7):1883–1896
54. Kahn LR (1952) Single-sideband transmission by envelope elimination and restoration. Proc IRE 40(7):803–806
55. Pedro JC, Garcia JA, Cabral PM (2007) Nonlinear Distortion Analysis of Polar Transmitters. In: IEEE MTT-S International Microwave Symposium Digest (IMS 2007), pp 957–960

56. Jinsung C, Dongsu K, Kang D, Kim B (2009) A polar transmitter with CMOS programmable hysteretic-controlled hybrid switching supply modulator for multistandard applications. IEEE Trans Microw Theory Techn 57(7):1675–1686
57. Feipeng W, Kimball DF, Popp JD, Yang AH, Lie DY, Asbeck PM, Larson LE (2006) An improved power-added efficiency 19 dbm hybrid envelope elimination and restoration power amplifier for 802.11 g WLAN applications. IEEE Trans Microw Theory Techn 54(12):4086–4099
58. Jinsung C, Kang D, Dongsu K, Kim B (2009) Optimized envelope tracking operation of doherty power amplifier for high efficiency over an extended dynamic range. IEEE Trans Microw Theory Techn 57(6):1508–1515
59. Dal Fabbro PA, Kayal M (2010) Linear CMOS RF power amplifiers for wireless applications. Springer, New York
60. Takahashi K, Yamanouchi S, Hirayama T, Kunihiro K (2008) An envelope tracking power amplifier using an adaptive biased envelope amplifier for WCDMA handsets. In: IEEE radio frequency integrated circuits symposium (RFIC 2008), pp 405–408
61. Myoungbo K, Jinseong J, Hassan M, Yan JJ, Kimball DF, Asbeck PM, Larson LE (2012) High efficiency wideband envelope tracking power amplifier with direct current sensing for LTE applications. In: IEEE topical conference on power amplifiers for wireless and radio applications (PAWR 2012), pp 41–44
62. Po-Hsing W, Yan L, Weibo H, Lopez J, Lie DYC, Liang TJ (2011) CMOS envelope tracking amplifier IC design for high-efficiency RF polar transmitters. In: IEEE international symposium on circuits and systems (ISCAS 2011), pp 197–200
63. Yan L, Lopez J, Schecht C, Ruili W, Lie DYC (2012) Design of high efficiency monolithic power amplifier with envelope-tracking and transistor resizing for broadband wireless applications. IEEE J Solid-State Circuits 47(9):2007–2018
64. Kang D, Byungjoon P, Chenxi Z, Kim D, Kim J, Cho Y, Sangsu J, Hadong J, Kim B (2012) A 34 % PAE, 26 dBm output power envelope-tracking CMOS power amplifier for 10 MHz BW LTE applications. In: IEEE MTT-S international microwave symposium digest (IMS 2012), pp 1–3
65. Hassan M, Olson C, Kovac D, Yan JJ, Nobbe D, Kelly D, Asbeck PM, Larson LE (2012) An envelope-tracking CMOS-SOS power amplifier with 50 % overall pae and 29.3 dbm output power for lte applications. In: IEEE compound semiconductor integrated circuit symposium (CSICS 2012), pp 1–4
66. Dal Fabbro PA, Meinen C, Kayal M, Kobayashi K, Watanabe Y (2006) A dynamic supply CMOS RF power amplifier for 2.4 GHz and 5.2 GHz frequency bands. In: IEEE radio frequency integrated circuits symposium (RFIC 2006), pp 169–172
67. Po-Chih W, Kai-Yi H, Yu-Fu K, Ming-Chong H, Chao-Hua L, Tzung-Ming C, Chia-Jun C, Ka-Un C, Ta-Hsun Y, Wen-Shan W, Ying-Hsi L, Lee C–C (2008) A 2.4 GHz +25 dBm P-1 dB linear power amplifier with dynamic bias control in a 65 nm CMOS process. In: Proceedings of european solid-state circuits conference (ESSCIRC 2008), pp 490–493
68. Xu Z, Gu QJ, Chang MCF (2011) A 100–117 GHz W-band CMOS power amplifier with on-chip adaptive biasing. IEEE Microw Wireless Compon Lett 21(10):547–549
69. Po-Chih W, Chia-Jun C, Wei-Ming C, Pei-Ju C, Chun-Cheng W, Chao-Hua L, Kai-Te C, Ming-Chong H, Yi-Ming C, Shih-Min L, Ka-Un C, Ying-His L, Lee C–C (2007) A 2.4 GHz fully integrated transmitter front end with +26.5 dBm on-chip CMOS power amplifier. In: IEEE radio frequency integrated circuits symposium (RFIC 2007), pp 263–266
70. Hyunji K, Bonhoon K, Songcheol H (2012) Highly efficient 24 GHz CMOS linear power amplifier with an adaptive bias circuit. In: IEEE asia-pacific microwave conference proceedings (APMC 2012), pp 7–9
71. Nai-Chung K, Jui-Chi K, Che-Chung K, Huei W (2011) K-band CMOS power amplifier with adaptive bias for enhancement in back-off efficiency. In: IEEE MTT-S international microwave symposium digest (IMS 2011), pp 1–4

72. Liu JYC, Tang A, Ning-Yi W, Gu QJ, Berenguer R, Hsieh-Hung H, Po-Yi W, Chewnpu J, Chang MCF (2011) A V-band self-healing power amplifier with adaptive feedback bias control in 65 nm CMOS. In: IEEE radio frequency integrated circuits symposium (RFIC 2011), pp 1–4
73. Chih-Yun L, Yi-Jan C, Heo D (2005) Impact of bias schemes on Doherty power amplifiers. In: IEEE international symposium on circuits and systems (ISCAS 2005), pp 212–215

Chapter 6
Power Amplifier Design

Abstract This chapter describes the design of a linear CMOS power amplifier. It begins with the description of a mathematical model that predicts the performance of the PA and then presents the PA's schematic design. In addition the chapter includes specific layout considerations for high frequency linear PAs. The design of the inductors implemented in PAs specific issues regarding layout are also discussed.

A Model for the Power Amplifier

The goal of any power amplifier is to obtain high output power levels and high PAE values along with maximum linearity. The process of optimizing a power amplifier is not simple and requires a starting point for the PA design. This chapter therefore proposes a model that takes into account effects such as PA biasing, current consumption, supply and breakdown voltages, inductor quality factors or the power gain in the P_{IN}-P_{OUT} and PAE-P_{IN} curves. The output of the model is precisely the starting point for the design.

Model Description

The model has been developed for fully integrated PAs and any type of linear (class A to class B) operation mode. Figure 6.1 shows the generic PA architecture that has been chosen for the model, whereas Fig. 6.2 shows its equivalent circuit. This model is a generic common-source stage with an output-matching circuit. The model also takes into consideration single or differential architectures and the use of cascode transistors. The architecture of the PA includes the drain inductor (L_D) and its parasitic resistor (R_D), and the output parasitic capacitor (C_O), which refers to the parasitics of the transistors and the drain inductor. It also comprises the output-matching network based on a series capacitor (C_S) and an inductor (L_P),

H. Solar Ruiz and R. Berenguer Pérez, *Linear CMOS RF Power Amplifiers*,
DOI: 10.1007/978-1-4614-8657-2_6,

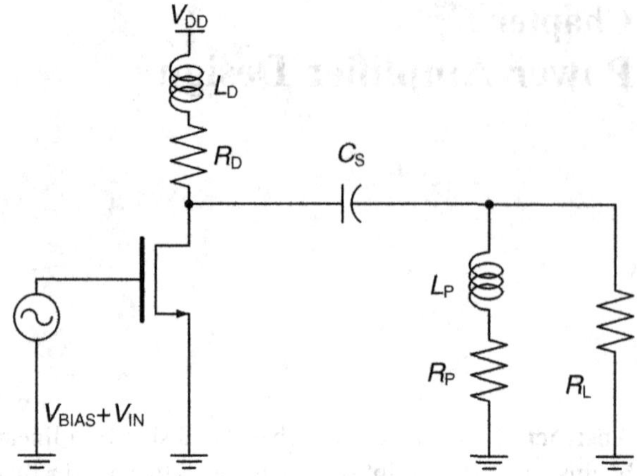

Fig. 6.1 Schematic circuit of the PA architecture

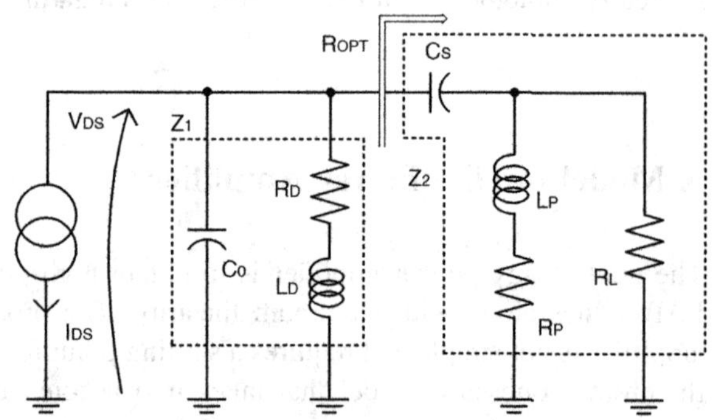

Fig. 6.2 Equivalent circuit of the PA architecture

with its parasitic resistor (R_P) [1]. The output-matching network converts R_L to the optimum load (R_{OPT}), which provides the best linearity to the PA [2].

The PA common-source transistor, along with a possible cascode transistor, is modeled as a current source with an output parasitic capacitor. This current source, as represented in Fig. 6.3 by the solid line, presents strong–weak nonlinear transconductance behavior with a maximum current level beyond which the current waveform suffers compression at input voltage overdriving conditions [2]. This strong–weak nonlinear transconductance model is more realistic than the strong nonlinear model, which is also shown in Fig. 6.3 by the dotted line.

Although less realistic, strong nonlinear transconductance will be used as a base for the construction of the strong–weak nonlinear transconductance and it will be of help in the calculations. The strong nonlinear model is described by (6.1).

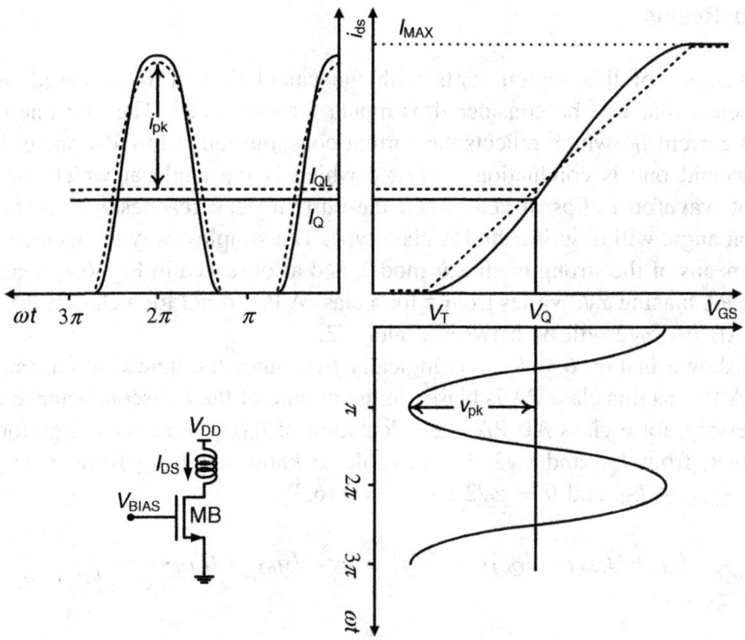

Fig. 6.3 Gate voltage and drain current waveforms for a class AB cascode PA. Normalized strong and strong–weak nonlinear models are shown for the transconductance

$$\begin{aligned}
&i_{ds} = k\left(V_Q + v_{pk}\cos(\theta)\right) - V_{TH} = I_{QL} + i_{pk}\cos(\theta) \\
&when \quad 0 < I_{QL} + i_{pk}\cos(\theta) < I_{MAX} \\
&i_{ds} = I_{MAX} \quad when \quad I_{QL} + i_{pk}\cos(\theta) > I_{MAX} \\
&i_{ds} = 0 \quad when \quad I_{QL} + i_{pk}\cos(\theta) < 0
\end{aligned} \tag{6.1}$$

In (6.1), I_{QL} is the linear quiescent current, V_{TH} the threshold voltage, θ is the phase and $k = 1$ for a normalized curve.

Strong–weak nonlinear transconductance is then described as a modification of linear normalized transconductance, as in (6.2).

$$\begin{aligned}
&i_{ds} = I_{MAX}\left(3\left(\frac{I_{QL} + i_{pk}\cos(\theta)}{I_{MAX}}\right)^2 - 2\left(\frac{I_{QL} + i_{pk}\cos(\theta)}{I_{MAX}}\right)^3\right) \\
&when \quad 0 < I_{QL} + i_{pk}\cos(\theta) < I_{MAX} \\
&i_{ds} = I_{MAX} \quad when \quad I_{QL} + i_{pk}\cos(\theta) > I_{MAX} \\
&i_{ds} = 0 \quad when \quad I_{QL} + i_{pk}\cos(\theta) < 0
\end{aligned} \tag{6.2}$$

The next subsections describe the proposed model and give the PAE and the P_{OUT} expressions for the linear and the nonlinear regions of generic class A to class B PAs.

Linear Region

The analysis of this region starts with the calculation of I_{MAX} based on two parameters that will be considered as inputs for the model. The first one is quiescent current I_Q, which reflects the current consumption at low PA power levels. The second one is conduction angle α_0, which is the angle at which the drain current waveform clips at zero when the current peak reaches I_{MAX}. The conduction angle will describe the PA class type. The simplest way to calculate I_{MAX} is by means of the strong nonlinear model, and as observed in Fig. 6.4, depending on the PA biasing $\alpha_0/2$ varies from π for a class A PA to $\pi/2$ for a class B PA. For a class AB PA $\alpha_0/2$ will be between π and $\pi/2$.

As shown in Fig. 6.4, I_{MAX} is logically two times the quiescent current for a class A PA, as this class PA is biased in the middle of the transconductance curve. Conversely, for a class AB PA, the calculation of I_{MAX} is not so straightforward. However, from I_{QL} and $\alpha_0/2$ it is possible to know the I_{MAX} from (6.1), when $i_{pk} = I_{MAX} - I_{QL}$ and $\theta = \alpha_0/2$ for $i_{ds} = 0$ (6.3).

$$I_{QL} + (I_{MAX} - I_{QL})\cos\left({}^{\alpha_0}/_2\right) = 0 \Rightarrow I_{MAX} = I_{QL} - \frac{I_{QL}}{\cos\left({}^{\alpha_0}/_2\right)} \tag{6.3}$$

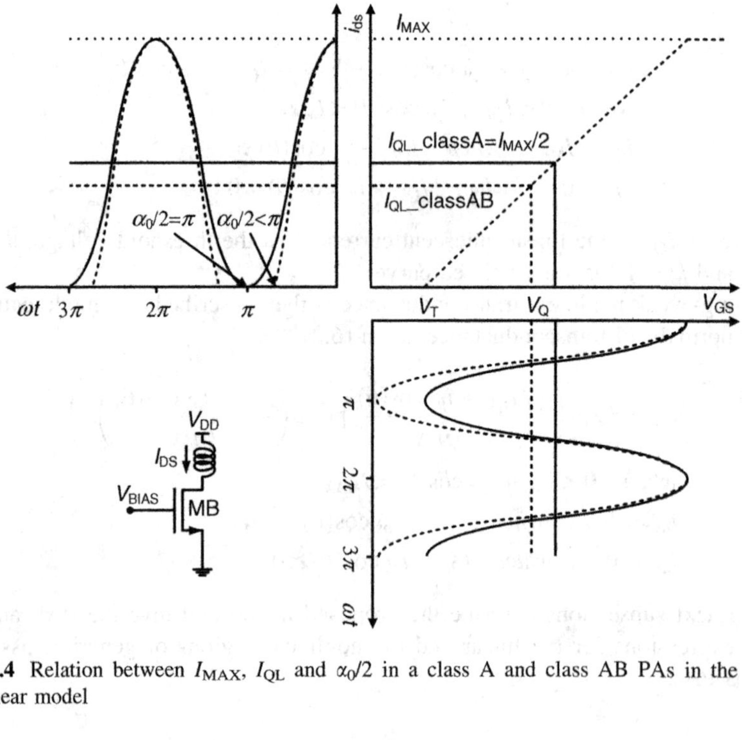

Fig. 6.4 Relation between I_{MAX}, I_{QL} and $\alpha_0/2$ in a class A and class AB PAs in the strong nonlinear model

On the other hand, as the input parameter is I_Q and not I_{QL}, I_{QL} must be calculated beforehand. The value of the linear quiescent current, I_{QL}, is obtained through I_Q, the quiescent current for the strong–weak model, from (6.4).

$$I_Q = I_{MAX}\left(3\left(\frac{I_{QL}}{I_{MAX}}\right)^2 - 2\left(\frac{I_{QL}}{I_{MAX}}\right)^3\right) \tag{6.4}$$

As α_0 is used as an input parameter, it is possible to substitute I_{MAX} in (6.4) with (6.3) so that I_{QL} can be easily calculated from (6.5).

$$I_Q = \frac{3I_{QL}}{1 - \frac{1}{\cos\left(\alpha_0/2\right)}} - \frac{2I_{QL}}{\left(1 - \frac{1}{\cos\left(\alpha_0/2\right)}\right)^2} \tag{6.5}$$

The situation shown in Fig. 6.4, i.e. the point at which the peak drain current i_{ds} reaches I_{MAX} just before starting to clip at its maximum, is the threshold between the linear and nonlinear region of a PA. It is considered the threshold because compression mainly takes place in the PA when the drain current waveform starts clipping at its maximum. Of course, in a class AB PA there is a previous clipping at zero in the drain current; however, the linearity of the PA is not greatly affected for that reason, as we will demonstrate with some results extracted from the model.

This consideration of the maximum linear point is based on the strong nonlinear model, but in the strong–weak nonlinear model some pre-compression is found in the drain current before reaching I_{MAX}. However, the value obtained with the strong nonlinear model is still the proper value for the optimum load calculation in the strong–weak nonlinear model.

Therefore, the aforementioned maximum linear point allows us to find the value of R_{OPT}, i.e. the PA load for the best linearity. R_{OPT} is chosen so that the drain current and the drain voltage waveforms reach their unclipped maximum values simultaneously, i.e. i_{ds} swings from 0 to I_{MAX} and v_{ds} swings from V_{KNEE} to $2V_{DD}$-V_{KNEE}. As mentioned in Chap. 3, the knee voltage is referred to the turn on voltage and significantly reduces the available RF swing because the voltage does not reach zero (and consequently it does not reach $2V_{DD}$ either). The V_{KNEE} is modeled as 20 % of the supply voltage [2]. In a PA that includes a cascode transistor the transistor stacking effect will be considered by increasing the V_{KNEE} two times that value.

The drain current and voltage waveforms at the maximum linear point are shown in Fig. 6.5 for i_{ds} and Fig. 6.6 for v_{ds} (dashed line). In this case, the waveforms have been obtained by assuming a broadband resistor load and the strong–weak nonlinear model so they show the abovementioned pre-compression near I_{MAX} and V_{KNEE}. In addition, as shown in Fig. 6.5, since it is a class AB PA, the drain current waveform clips at zero before clipping at I_{MAX}. This clipping at zero is also reflected in Fig. 6.6 for the clipped aspect of drain voltage waveform at its maximum.

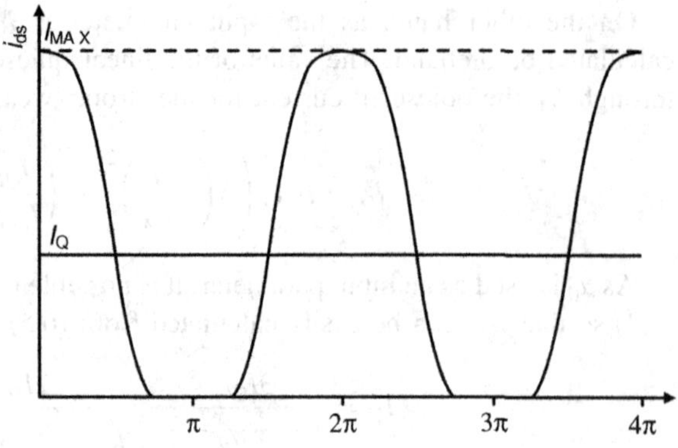

Fig. 6.5 Class AB PA drain current waveform at its maximum current swing before clipping

Therefore, in order to get the simultaneous maximum swings for i_{ds} and v_{ds}, the output load (R_L) of the PA must be transformed to the optimum load (R_{OPT}), which can be expressed by (6.6).

$$R_{OPT} = \frac{V_{DD} - V_{KNEE}}{I_{MAX}/2} \tag{6.6}$$

It must be noted that the value of R_{OPT} that results from (6.6) is correct only if the drain inductor, L_D, is considered ideal. However, as it is an integrated low Q inductor, the effect of its parasitic resistor, R_D, must be taken into account. As shown by the solid line in Fig. 6.6, the DC value of this parasitic resistor ($R_{D,DC}$) reduces the effective supply voltage from V_{DD} to V'_{DD}, and consequently the maximum unclipped value of the transistor drain voltage is reduced. In fact, the L_D parasitic resistor greatly affects the performance of a PA due to the high currents flowing through it. The effective supply voltage is reduced according to (6.7):

$$V'_{DD} = V_{DD} - I_{DC,0} \cdot R_{D,DC}. \tag{6.7}$$

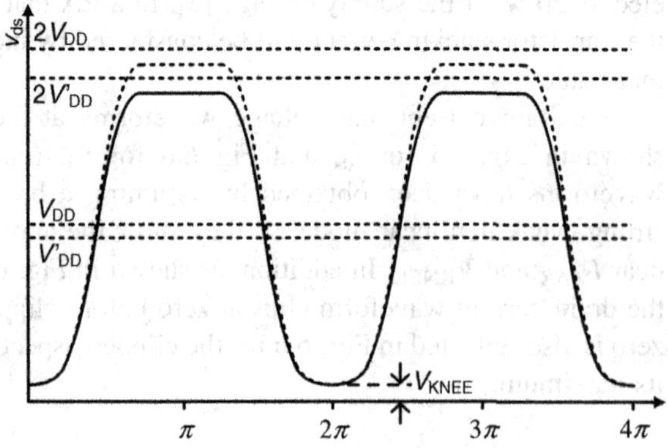

Fig. 6.6 Class AB PA drain voltage waveform at its maximum voltage swing before clipping. The swing of the voltage waveform is reduced from V_{DD} (*dashed line*) to V'_{DD} (*solid line*) due to the effect of the parasitic resistor of the drain inductor

The value of the $I_{DC,0}$ current is obtained from (6.17) precisely when the drain current reaches the maximum linear point. Consequently, the value of i_{ds} that is introduced in (6.17) is obtained from (6.2) when $i_{pk} = I_{MAX} - I_{QL}$.

Moreover, the high frequency value of R_D ($R_{D,AC}$) also affects the value of R_{OPT} because it makes the real part of impedance Z_1, shown in Fig. 6.2, be finite. Consequently, the optimum load must finally fulfill (6.8).

$$R_{OPT}//\mathrm{Re}\{Z_1\} = \frac{V_{DD}' - V_{KNEE}}{I_{MAX}/2}. \tag{6.8}$$

Z_1 is given by (6.9):

$$Z_1 = (R_{D,AC} + j\omega L_D)//\frac{1}{j\omega C_0} \tag{6.9}$$

The real part of Z_1 is obtained from (6.10), while the value of L_D (in charge of resonating out capacitance C_0) is given by (6.11):

$$\mathrm{Re}\{Z_1\} = \frac{\omega_0^2 L_D^2 + R_{D,AC}^2}{R_{D,AC}} \tag{6.10}$$

$$\mathrm{Im}\{Z_1\} = 0 \Rightarrow L_D = \frac{1 + \sqrt{1 - (2\omega_0 R_{D,AC} C_0)^2}}{2\omega_0^2 C_0} \tag{6.11}$$

As it has already been mentioned, R_{OPT} is obtained by means of an impedance downconversion network formed by L_P and C_S that transforms R_L to R_{OPT}. Ideally, the impedance downconversion network is lossless. However, an integrated network again makes use of low Q inductors so that the extra losses must be considered in the design. Specifically, the model includes the effect of the high frequency parasitic resistor of the L_P inductor ($R_{P,AC}$), which reduces the actual transmitted power to R_L and must be taken into account in calculations of P_{OUT}. The values of L_P and C_S are (6.13) and (6.14) when impedance Z_2, in Fig. 6.2, is matched to R_{OPT}.

$$Z_2 = \left[(R_{P,AC} + j\omega_0 L_P)//R_L\right] + \frac{1}{j\omega_0 C_S} \tag{6.12}$$

$$\mathrm{Re}\{Z_2\} = R_{OPT} \Rightarrow L_P = \frac{1}{\omega_0}\sqrt{\frac{(R_{P,AC} + R_L)\left[(R_{P,AC} + R_L)R_{OPT} - R_{P,AC}R_L\right]}{R_L - R_{OPT}}} \tag{6.13}$$

$$\mathrm{Im}\{Z_2\} = 0 \Rightarrow C_S = \frac{(\omega_0 L_P)^2 + (R_{P,AC} + R_L)^2}{\omega_0^2 L_P R_L^2}. \tag{6.14}$$

Now, in order to obtain P_{OUT} and PAE it is necessary to calculate the DC and the fundamental component of the drain current. Two different situations must be

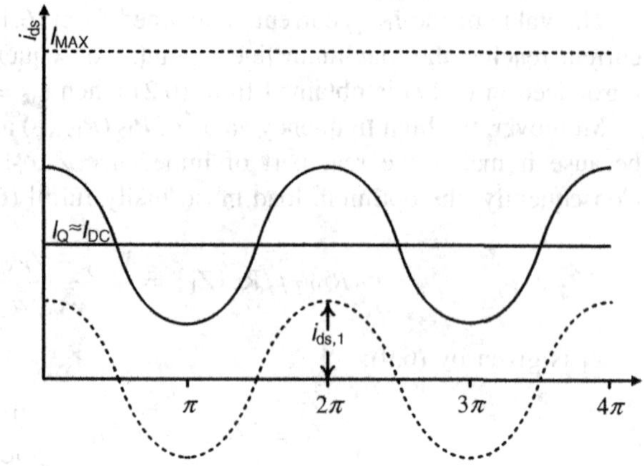

Fig. 6.7 Drain current waveform for the PA in the linear region. It corresponds to the current levels before zero clipping. The DC and fundamental components of the current are also shown

distinguished in the linear region for a generic class AB PA. Both situations are illustrated in Figs. 6.7 and 6.8.

Figure 6.7 is an example of current amplitude in which the current has not yet reached zero clipping. In this case the DC and the fundamental component of the drain current are obtained from (6.15) and (6.16) for different i_{pk} values in (6.2). As observed in Fig. 6.7, the DC current is almost constant and similar to I_Q.

$$I_{DC} = \frac{1}{2\pi}\int_0^{2\pi} i_{ds} d\theta \tag{6.15}$$

$$i_{ds,1} = \frac{1}{\pi}\int_0^{2\pi} i_{ds}\cos(\theta) d\theta \tag{6.16}$$

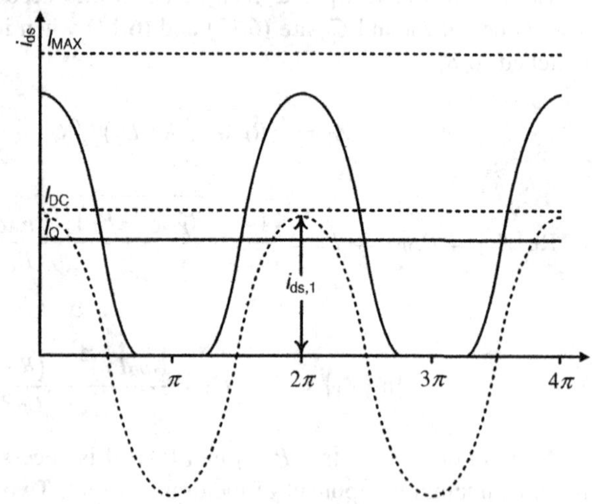

Fig. 6.8 Drain current waveforms for the PA in the linear region. It corresponds to the current levels after zero clipping but before reaching its maximum. The DC and fundamental components of the current are also shown

On the other hand, when the drain current is clipped at zero, the situation is the one observed in Fig. 6.8. The DC and fundamental components of the current are obtained from (6.17) and (6.18). In such a case the I_{DC} starts increasing above I_Q.

$$I_{DC} = \frac{1}{2\pi}\int_0^{\alpha} i_{ds} d\theta \tag{6.17}$$

$$i_{ds,1} = \frac{1}{\pi}\int_0^{\alpha} i_{ds}\cos(\theta) d\theta \tag{6.18}$$

Both expressions are similar to (6.15) and (6.16) but, in this case, the integral limits are from 0 to α, the current zero-clipping angle. As observed in Fig. 6.9, angle α can be obtained by means of the strong nonlinear model when $i_{ds} = 0$ while varying the i_{pk} value (6.19).

$$\alpha/2 = \arccos\left(\frac{-I_{QL}}{i_{pk}}\right) \tag{6.19}$$

Angle α varies for different i_{pk} values up to $i_{pk} = I_{MAX} - I_{QL}$, i.e. until i_{pk} reaches the maximum linear point and where $\alpha/2 = \alpha_0/2$. Values higher than

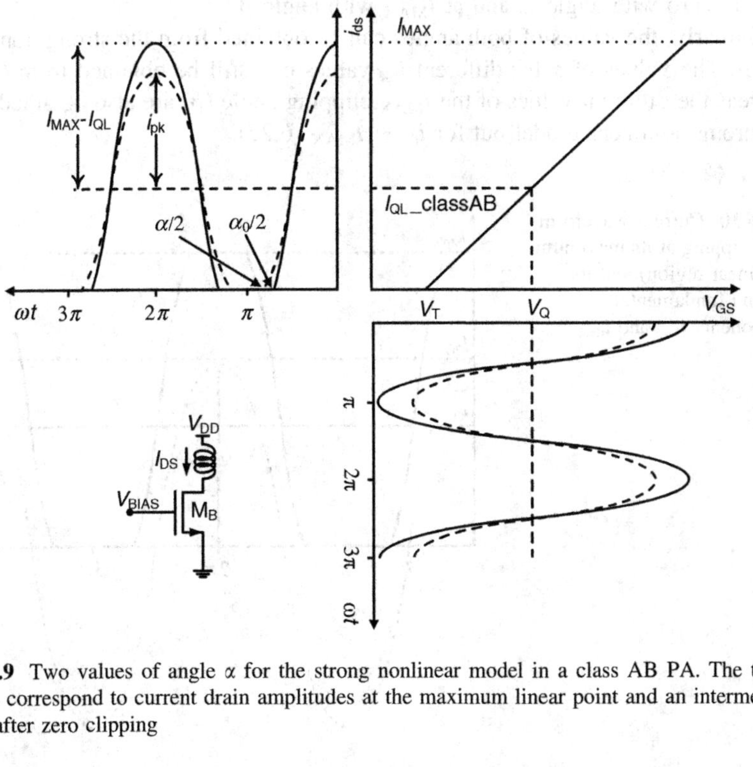

Fig. 6.9 Two values of angle α for the strong nonlinear model in a class AB PA. The two α values correspond to current drain amplitudes at the maximum linear point and an intermediate level after zero clipping

$I_{MAX} - I_{QL}$ for i_{pk} also modify angle α but also lead the PA to the nonlinear region. This region is discussed in the next section.

Nonlinear Region

The overdriving conditions, i.e. the nonlinear region, allow P_{OUT} and PAE values to go beyond the ones found in the linear region due to the increase of the voltage and current fundamental components [2]. Figure 6.10 shows the drain current waveform of a PA in an overdriven situation and its fundamental ($i_{ds,1}$) and DC (I_{DC}) components.

The fundamental and DC components are calculated in a manner similar to the linear region using (6.20) and (6.21). The only difference is that now it must be considered that the current also clips at I_{MAX}, as shown in Fig. 6.10.

$$I_{DC} = \frac{1}{2\pi}\left(\int_{\beta}^{\alpha} i_{ds} d\theta + I_{MAX}\beta\right) \tag{6.20}$$

$$i_{ds,1} = \frac{1}{\pi}\left(\int_{0}^{\beta} I_{MAX}\cos(\theta)d\theta + \int_{\beta}^{\alpha} i_{ds}\cos(\theta)d\theta\right) \tag{6.21}$$

As observed in Fig. 6.11, the drain current waveform in this region now clips both: at zero with angle α, and at I_{MAX} with angle β.

Similarly, the values of both angles can be obtained from the strong nonlinear model. The values of α for different i_{pk} values can still be obtained from (6.19), whereas the different values of the I_{MAX} clipping angle (β) are also obtained from the strong nonlinear model but for $i_{ds} = I_{MAX}$ (6.22).

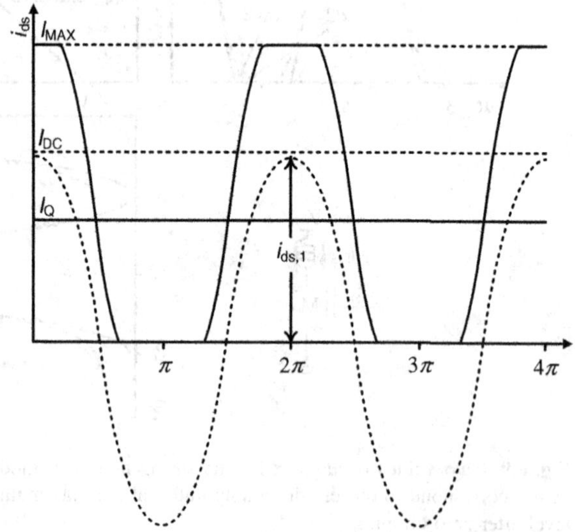

Fig. 6.10 Current waveform after clipping at its maximum (nonlinear region) and its DC and fundamental components I_{DC} and $i_{ds,1}$

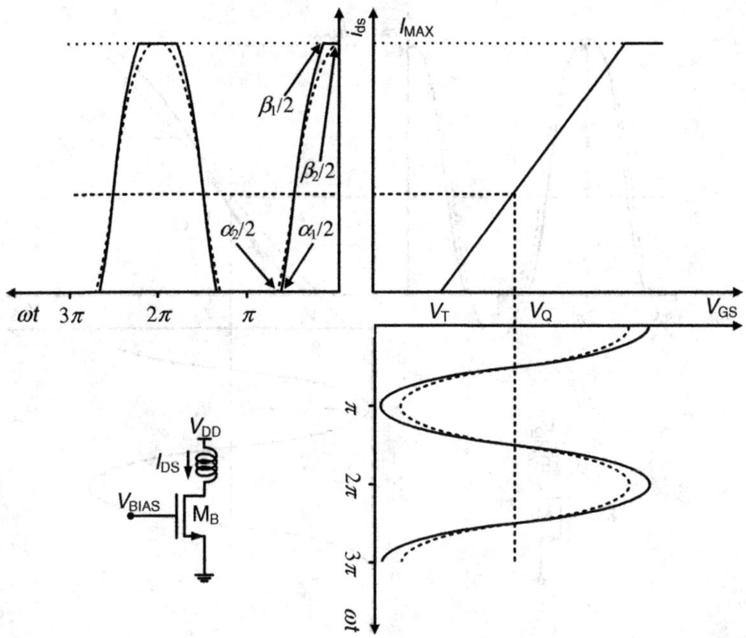

Fig. 6.11 Two pairs of values of angles α and β for the strong nonlinear model in a class AB PA. As the PA is in the nonlinear region, the two pairs of values correspond to current drain amplitudes that shows clipping both at zero and I_{MAX} after zero clipping

$$\beta/2 = \arccos\left(\frac{I_{MAX} - I_{QL}}{i_{pk}}\right) \tag{6.22}$$

Now, based on the previous equations, it is possible to calculate the output power, the drain efficiency and the PAE for the different regions.

Output Power, Drain Efficiency and PAE

With regard to output power, when (6.13) and (6.14) are fulfilled, P_{OUT} is expressed by (6.23).

$$P_{OUT} = \frac{i_{ds,1}^2 (R_{OPT}//\mathrm{Re}\{Z_1\})^2 \left(1 + \frac{1}{(\omega_0 C_S R_{OPT})^2}\right)}{2R_L} \tag{6.23}$$

Consequently, the drain efficiency is (6.24).

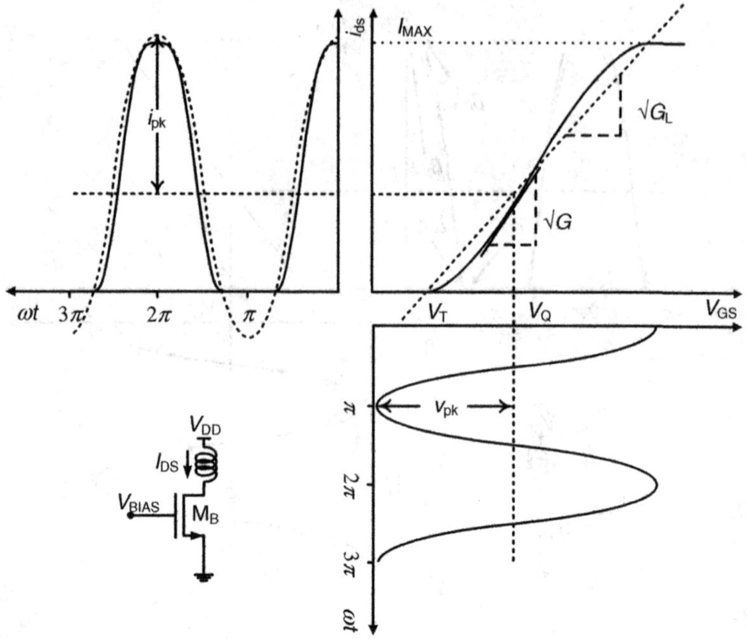

Fig. 6.12 Gate voltage and drain current waveforms for a class AB cascode PA. Normalized linear and strong–weak nonlinear models are shown for the transconductance

$$\eta(\%) = 100\frac{P_{OUT}}{P_{DC}} = 100\frac{\frac{i_{ds,1}^2\left(R_{OPT}//\mathrm{Re}\{Z_1\}\right)^2\left(1+\frac{1}{(\omega_0 C_S R_{OPT})^2}\right)}{2R_L}}{V_{DC}I_{DC}} \tag{6.24}$$

In order to obtain values for the PAE, it is first necessary to establish a connection between i_{pk} and P_{IN}. Once the connection is made, by providing different values for P_{IN}, different values for i_{pk} are obtained and hence for i_{ds}. This is achieved in an indirect way by giving different values to P_{IN} and assuming linear transconductance, i.e. similar to strong nonlinear transconductance but with no clipping limits, as shown in Fig. 6.12 by the dashed line.

Where G_L is the linear gain of the ideal transconductance, the linear output power can be obtained by (6.25), which allows peak drain current values to be obtained from (6.26).

$$P_{OUT,L} = G_L P_{IN} \tag{6.25}$$

$$i_{pk} = \sqrt{\frac{2R_L P_{OUT,L}}{\left(R_{OPT}//\mathrm{Re}\{Z_1\}\right)^2\left(1+\frac{1}{(\omega_0 C_S R_{OPT})^2}\right)}} \tag{6.26}$$

The resulting values of i_{pk} are used for calculations in the weak-strong model of transconductance in order to obtain i_{ds} and hence I_{DC} and $i_{ds,1}$. With these current values it is then possible to calculate P_{OUT} values in either the linear or nonlinear region.

On the other hand, linear gain G_L cannot be used directly as an input parameter because, as shown in Fig. 6.12, it does not have to coincide with the gain of the strong-weak nonlinear model (G), i.e. the slope of the linear transconductance and the strong–weak nonlinear slope at the specific class polarization is not the same. However, as it is desirable to establish gain G as an input parameter, it is necessary to perform a gain correction by means of angle α_0 and (6.27). Therefore, if gain G is used as an input parameter, G_L is adjusted. For example, for a class A PA and low power levels, the value for G is 3.5 dB higher than G_L due to the difference in slopes of the linear and strong–weak nonlinear transconductances in the middle of the curve. Therefore, if gain G of class A PAs is set to 25 dB as an input parameter, G_L must be set to 21.5 dB.

$$G(dB) = G_L(dB) + 20\log\left(6\left(\frac{1}{1-\cos\left(\alpha_0/2\right)} - \left(\frac{1}{1-\cos\left(\alpha_0/2\right)}\right)^2\right)\right) \tag{6.27}$$

The PAE is calculated using (6.28) for different P_{IN} values.

$$PAE(\%) = 100\frac{P_{OUT} - P_{IN}}{P_{DC}} \tag{6.28}$$

Finally, Fig. 6.13 shows the flowchart with the parameter dependencies from the input to the output parameters and the equations that should be applied for the specific parameter calculations.

Model-Based Analyses

The proposed model allows the quantification of PA effects such as current consumption, supply voltage, transistor stress level, inductor quality factors, gain and PA biasing (from class A to class B). Therefore, different conclusions can be extracted by applying the model to different cases. Furthermore, it is possible to establish fast comparisons between different PA configurations in order to choose what is optimal for starting a new design.

The following sections show some examples of the model's capabilities as a helpful tool for PA design.

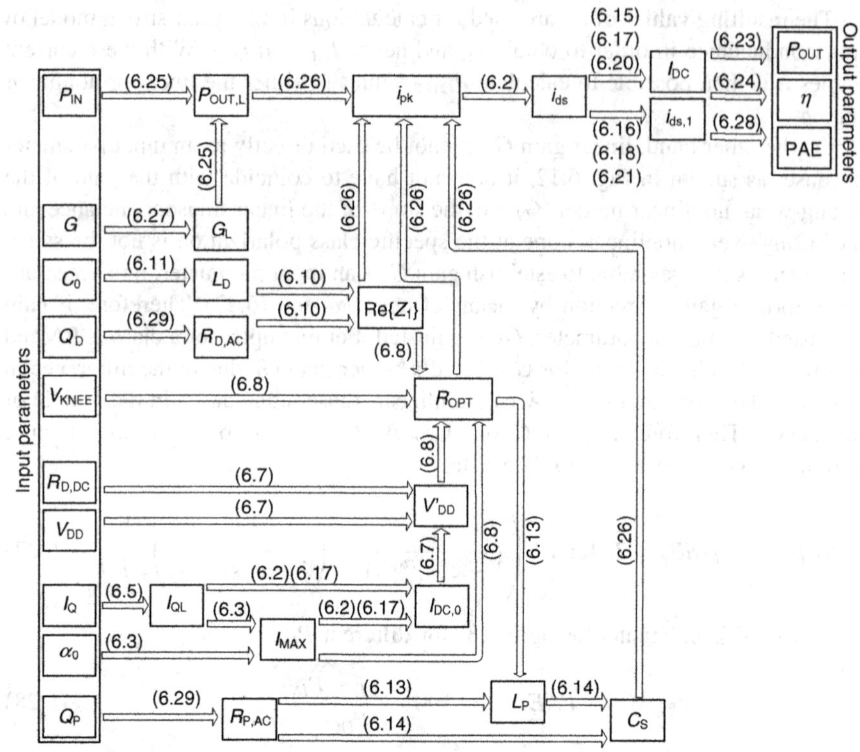

Fig. 6.13 Flowchart of the parameter dependencies and the equations that should be applied to parameter calculation

Effect of the Cascode Transistor

In Figs. 6.14 and 6.15 compare a light class AB ($\alpha_0 = 1.7\pi$) cascode PA and a class AB common-source PA with the same biasing. The comparison considers the effect of the knee and the breakdown voltages on the P_{OUT} and PAE parameters.

The design parameters used in the comparison are the following: a gain of 25 dB, a knee voltage of 0.36 V (this value is 0.72 V to consider the effect of the cascode transistor) and an I_Q of ~300 mA. A driver stage is added with half the current consumption, so that the 25 dB gain is a realistic value (of course, the effect of the driver stage is included in efficiency calculations). The $R_{D,DC}$ is 1 Ω and the quality factors are 10 for L_D and L_P. The high frequency resistor (R_{AC}) is calculated from the Q of the inductors at the frequency of interest using (6.29). A frequency of 5 GHz has been set.

$$R_{AC} = \frac{\omega_0 L}{Q} \tag{6.29}$$

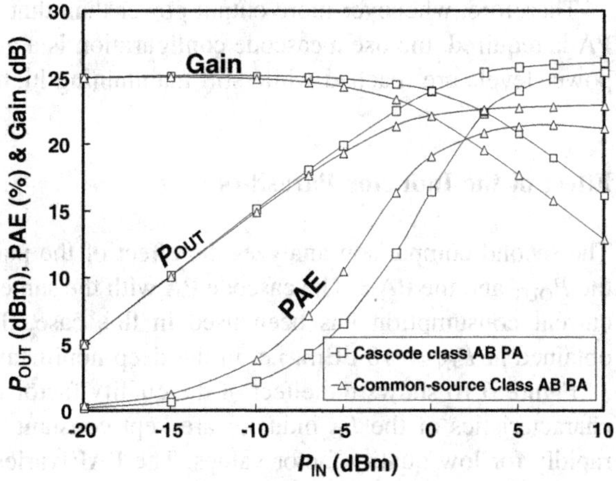

Fig. 6.14 A comparison of the P_{SAT}, gain and PAE parameters of a cascode and a common-source class AB PA

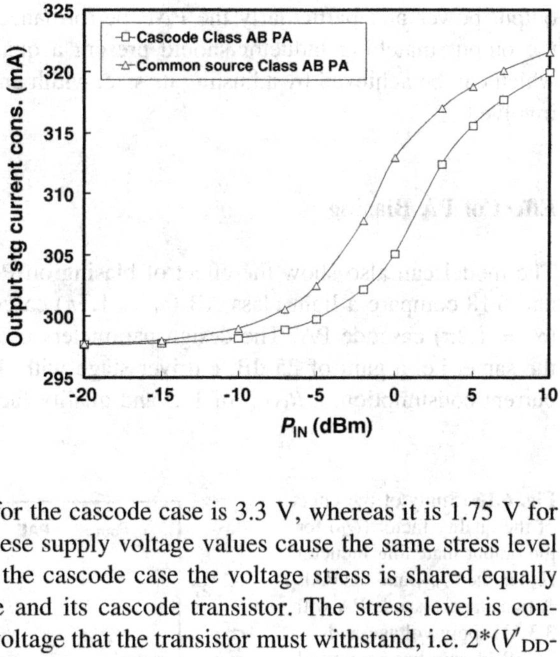

Fig. 6.15 A comparison of the current consumption levels of a cascode and a common-source class AB PA

The supply voltage used for the cascode case is 3.3 V, whereas it is 1.75 V for the common-source case. These supply voltage values cause the same stress level in the transistors because in the cascode case the voltage stress is shared equally between the common-source and its cascode transistor. The stress level is considered the maximum drain voltage that the transistor must withstand, i.e. $2*(V'_{DD}-V_{KNEE})$.

As can be seen in Fig. 6.14, the output power of the cascode configuration is more than 3 dB greater than the common-source configuration at P_{SAT}. On the other hand, the PAE is higher for the common-source PA at low power values but at P_{1dB}, which is considered the limit between the linear and nonlinear region, the PAE is higher for the cascode PA, 16.5 % versus 12.5 % for the common-source PA. Finally, the maximum PAE for the cascode PA is 28 % whereas it is 24 % for the common-source PA.

Therefore, whenever more output power than that offered by a common-source PA is required, the use a cascode configuration is a good choice, as greater output power levels are reached while still maintaining high PAE values.

Effect of the Inductor Parasitics

The second comparison analyzes the effect of the parasitics of the L_P inductor in the P_{OUT} and the PAE. The cascode PA with the same previous supply voltage and current consumption has been used in this case. The P_{OUT} values have been obtained at $P_{IN} = 10$ dBm, i.e. in the deep nonlinear region.

Figure 6.16 shows the effect of the quality factor for the L_P inductor while the characteristics of the L_D inductor are kept constant. The P_{OUT} and the PAE fall rapidly for low quality factor values. The PAE varies from 28 to 19 % when the quality factor takes values from 14 to 4. Likewise, the P_{OUT} degrades almost by 2 dB. As observed, when the quality factor of the L_P inductor falls below 8, the output power and particularly the PAE performance degrade rapidly. Therefore, the output matching inductor should present a quality factor above that value, which can be achieved by adjusting its size, width and the number of metal layers involved.

Effect of PA Biasing

The model can also show the effect of biasing on PA performance. Figures 6.17 and 6.18 compare a light class AB ($\alpha_0 = 1.7\pi$) cascode PA and a deep class AB ($\alpha_0 = 1.2\pi$) cascode PA. The design parameters used within the comparison are the same, i.e. a gain of 25 dB, a driver stage with the same biasing and half the current consumption, a $R_{D,DC}$ of 1 Ω and quality factors of 10 for L_D and L_P. As

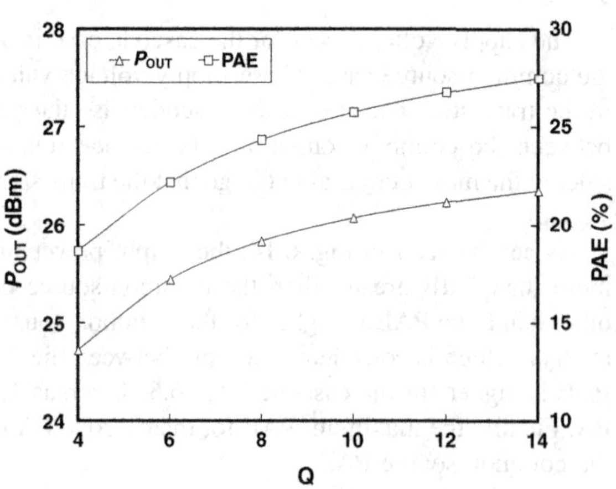

Fig. 6.16 Study of the effect of the quality factor (Q_P) for the output matching inductor (L_P) in the P_{OUT} and the PAE. A two stage cascode PA with 3.3 V supply voltage and 25 dB of gain has been used

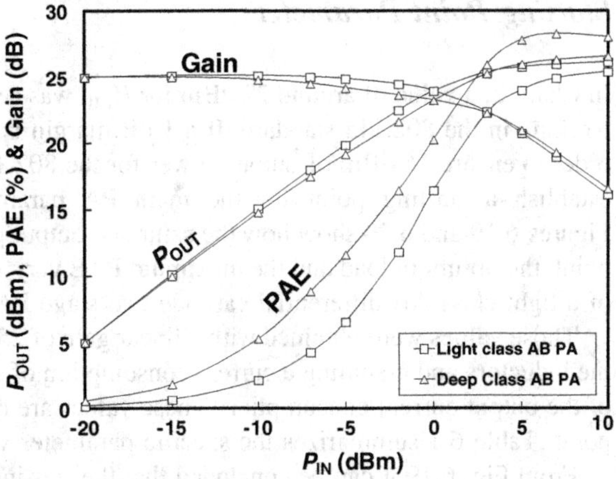

Fig. 6.17 A comparison of the P_{SAT}, gain and PAE parameters for a cascode light and a deep class AB PA

observed in Fig. 6.18 the I_Q has been chosen so that the PA shows the same current consumption at the highest output power levels.

It can be observed that the deep polarized class AB PA shows higher efficiency because it offers similar output power levels for less current consumption. However, this comes at the price of worse linearity as illustrated by the P_{IN}-P_{OUT} curve of Fig. 6.17. The deep polarized class AB shows precompression that degrades the P_{1dB}.

From Fig. 6.18 it can be also concluded that the layout design of the PA metal layers must be done carefully because deep class AB PAs have higher current consumption variations relative to input power. The width of the layout metal layer must consider the current consumption at the highest output power levels and not in the linear PA range. However, as it is also shown in Fig. 6.18, a light class AB PAs has less current variation relative to the input power.

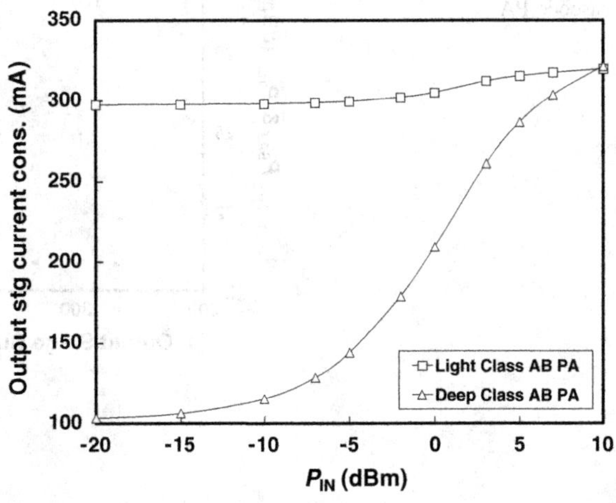

Fig. 6.18 A comparison of the current consumption levels for a cascode light and deep class AB PA

Starting Point Parameters

In Chap. 2, a value of around 22 dBm for P_{1dB} was set as a proper choice for a PA working in the 802.11a standard. If a 1 dB margin is chosen ($P_{1dB} = 23$ dBm) in order to ensure 16 dBm of output power for the 802.11a standard, it is possible to establish a starting point for the main PA parameters by using the model. Figures 6.19 and 6.20 show how the saturated output power, the 1 dB compression point, the optimum load and the maximum PAE vary with the current consumption of a light class AB differential cascode two stage PA.

Those values were obtained with a linear gain of 25 dB, quality factors of 10 for the inductors and assuming a current consumption of a first driver stage that is half of the output current consumption. These values are reasonable ones for a starting point. Table 6.1 summarizes the specific parameter values applied to the model.

From Fig. 6.19 it can be concluded that the starting point of the PA complying with the P_{SAT}/P_{1dB} can be achieved with a cascode differential two stage PA with a 3.3 V supply voltage, quality factors of 10 for the inductors and a current consumption close to 300 mA for the output stage and 150 mA for the driver stage. In addition, with those parameters it is possible to have a maximum PAE of 27 %. Finally, the load must downconverted from differential 100 Ω to differential 30 Ω.

The model has been used to determine a starting point for the design where parameters can be dynamically changed to show the effect of variation for any of them. For example, if a quality factor for an inductor is different, the model will easily show the impact of the new inductor Q.

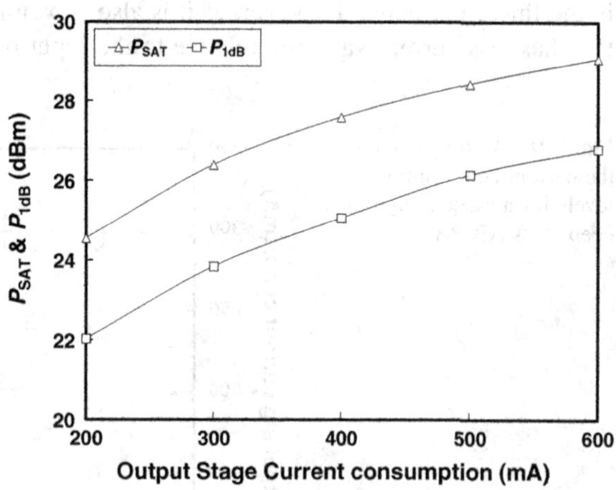

Fig. 6.19 P_{SAT} and P_{1dB} versus current consumption obtained from the model for a 25 dB gain, two stage cascode PA

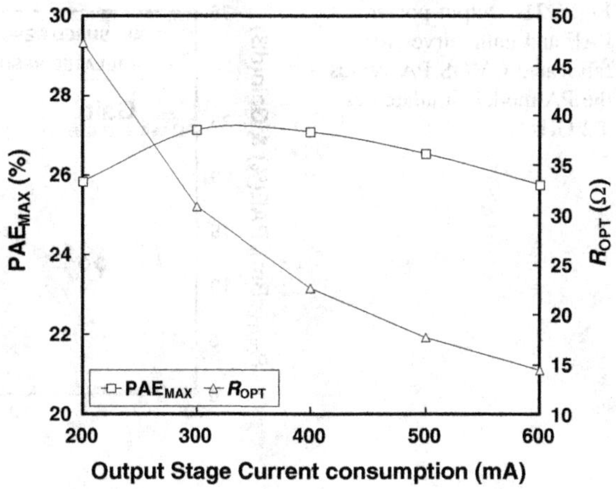

Fig. 6.20 Optimum load and maximum PAE versus current consumption obtained from the model for a 25 dB gain, two stage cascode PA

Table 6.1 Parameter values used in the model

Parameter name	Value	Units
V_{DD}	3.3	V
Output stage I_{DC}	200–600	mA
Driver stage I_{DC}	100–300	mA
Q_P	10	–
Q_D	10	–
$R_{D,DC}$	1	Ω
Gain	25	dB

Measurement Comparisons

Some results of a fabricated PA are now presented in order to show that the model fits well with measurements. Figure 6.21 illustrates the output power, gain and PAE values for a 3.3 V supply voltage and Fig. 6.22 shows the current consumption of the driver and output stages.

Figures 6.21 and 6.22 compare the measured curves with the ones provided by the model, showing good agreement and demonstrating the accuracy of the model. The main deviations in terms of power are observed at the highest power levels. Other nonlinear sources apart from the nonlinear behavior of the transistor current source explain these deviations. However, as these errors are kept low, no additional nonlinear sources are required in the model and thus the model is easy to handle.

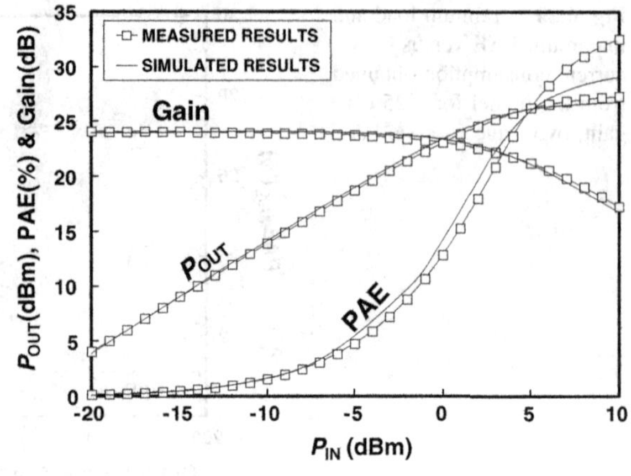

Fig. 6.21 Output power, PAE and gain curves for a fabricated CMOS PA versus the PA model simulated at 4.2 GHz

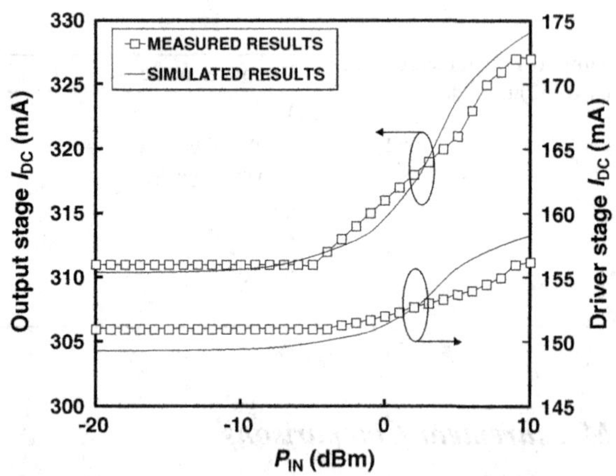

Fig. 6.22 Current consumption of a fabricated CMOS PA versus the PA model simulated at 4.2 GHz

Power Amplifier Design

The description of each stage of the PA architecture is explained in the next subsections, along with the design flow that has been followed.

Schematic Circuit Description

The PA illustrated in Fig. 6.23 shows a common architecture for fully integrated linear power amplifiers. The circuit consists of two stages in a cascode differential configuration. The use of two stages ensures the desired power gain for the PA, and the cascode topology improves linearity, as it is possible to have higher voltage swings at the output compared to a single transistor, due to the limitations imposed by the phenomena related to oxide breakdown. Both stages are connected through an interstage matching network based on a series resistor and a series capacitor pair. The output matching network downconverts the output load to the optimum load. Finally, the input network matches the input of the PA to 50 Ω.

Design Flow

The design flow that must be followed for PA design and the influence that the main parameters have on PA performance are now described. The methodology allows the optimization of any linear PA designed for high frequency applications, as is the case for the standards in the U-NII band, for example.

It has already been shown how the previously described model helps the designer set an initial value for the main PA parameters, leading to a value of 300 mA for the current consumption of the output stage with a 3.3 V supply voltage.

Once a value for the current consumption has been obtained, the size of the output stage transistors can be determined. A simulation process must be

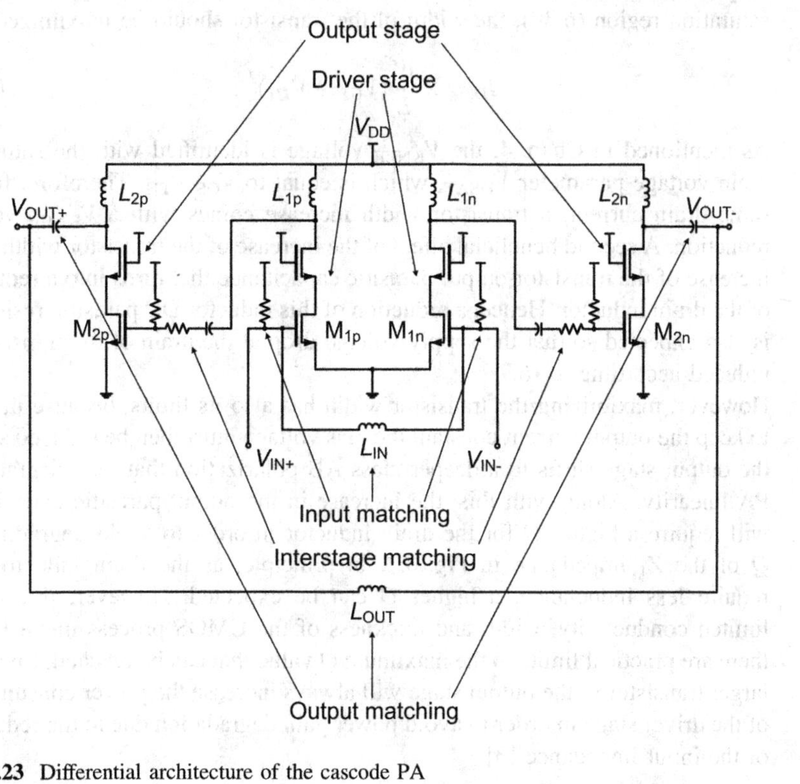

Fig. 6.23 Differential architecture of the cascode PA

performed by modifying the transistor size, the bias and the output load value until an optimum transistor width is reached in terms of linearity, output power and efficiency. In this optimization process, the performance of the drain and output inductors is essential as they affect both efficiency and linearity. The main parameters affecting overall PA performance that must be considered in the optimization process are the following:

1. The output matching network: parallel inductor and series capacitors.
   Starting from the value obtained from the model, the optimum output load must be found for the output stage of the PA. That value will fix the inductor and capacitor values of the output downconversion network.
2. The common-source and cascode transistor size, the gate voltage bias and the drain inductor value of the output stage.
   These parameters are related. The final size of the transistors is related to the voltage bias of the common-source gates because both affect current consumption. Similarly, the size of the transistors affects the output parasitic capacitance, which changes the inductance of the drain inductors, as they resonate out the parasitic capacitance. Furthermore, the transistor size affects the parasitics that result from the transistor interconnections made during the layout process, which will again change the optimum drain inductor value.
   In principle, when attending to the transistor drain current equation in the saturation region (6.30), the width of the transistor should be maximized.

$$I_D = K\frac{W}{L}(V_{GS} - V_{TH})^2 \tag{6.30}$$

   As mentioned in Chap. 4, the V_{KNEE} voltage is identified with the saturation drain voltage parameter V_{DSAT}, which is equal to V_{GS}-V_{TH}. Therefore, for the same drain current, a transistor width increase comes with a V_{KNEE} voltage reduction. A second beneficial effect of the increase of the transistor width is the increase of the transistor output parasitic capacitance that turns into a reduction of the drain inductor. Hence, a reduction of this inductor DC parasitic resistance is also expected so that the supply voltage drop at the drain of the transistor is reduced according to (6.7).
   However, maximizing the transistor width has also its limits, because in order to keep the output current constant the bias voltage must then be reduced so that the output stage shifts to a deeper class AB polarization that may degrade the PA linearity. Along with this, the increase in the output parasitic capacitance will require a higher Q for the drain inductor in order to avoid degrading the Q of the Z_1 impedance in Fig. 6.2. In principle, as the drain inductor will require less inductance, a higher Q can be expected. However, due to the limited conductivity, width and thickness of the CMOS process metal layers, there are practical limits to the maximum Q value that can be reached. Finally, a larger transistor in the output stage will always increase the power consumption of the driver stage in order to avoid power gain degradation due to the reduction of the input impedance [3].

On the other hand, it must be also noted that there is a minimum transistor size, as the gates of the transistor have a maximum current density that must not be exceeded.

3. The common-source and cascode transistor size, the gate voltage bias and the drain inductor value of the driver stage.
   The issues explained for the output stage must be also addressed for the driver stage. This stage is designed to provide enough gain for the overall PA. However, its current consumption, transistor size and drain inductor value should be carefully chosen, as this stage affects the efficiency, linearity and output power values of the overall PA.
4. The input matching network.
   The input matching network, which is composed of the resistor divider and the input inductor, is considered at the end of the optimization process.

Output Matching Network

The output matching network must be designed as an impedance downconverter from the 50 Ω output load to the optimum load. Figures 6.24, 6.25 and 6.26 illustrate common examples of single impedance downconversion networks at RF frequencies. These examples are suitable for integration as they present few components, which translates into lower losses if compared to more complex networks.

The first topology, the low pass L-match network, which appears in Fig. 6.24, consists of a parallel capacitor and a series inductor. Although it is a simple topology, it has a drawback when it is applied to a differential architecture because in this case two integrated inductors are necessary. This is not a minor drawback, as inductors are the largest components in any RF IC layout.

On the other hand, the second topology, the high pass L-match network, shown in Fig. 6.25, has several advantages when compared to the previous one. This topology permits the use of a unique inductor for differential architectures when a

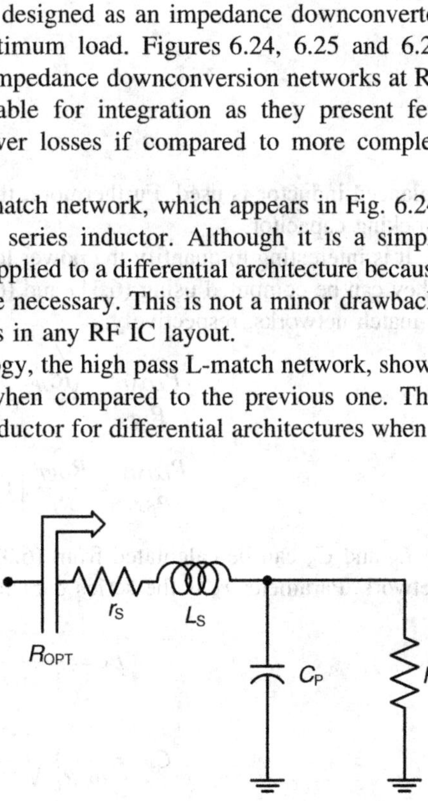

Fig. 6.24 A low pass L-match impedance downconverter

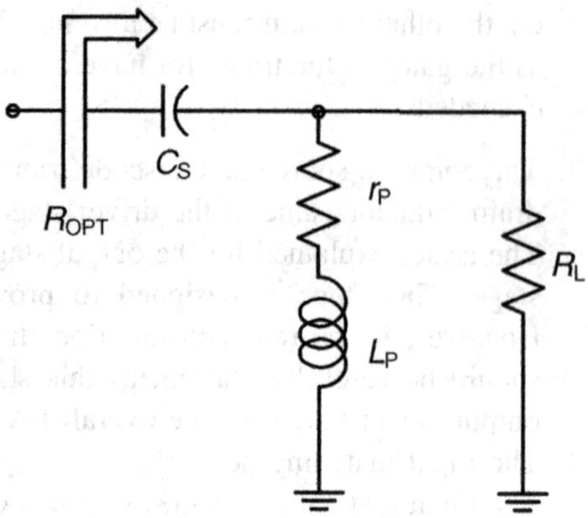

Fig. 6.25 A high pass L-match impedance downconverter

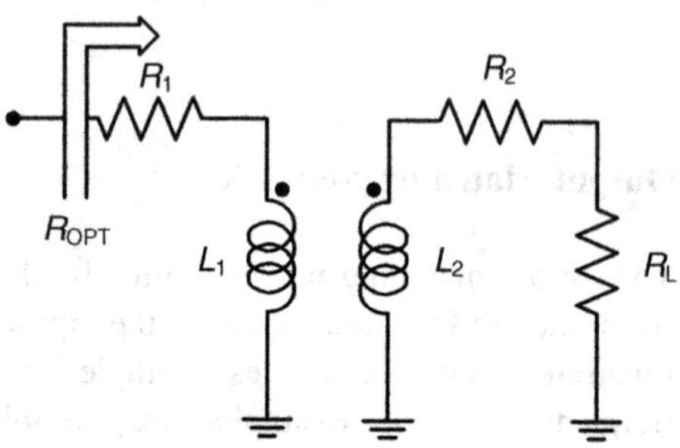

Fig. 6.26 A transformer based impedance downconverter

balanced inductor is used. Furthermore, the series capacitor can also act as a DC blocking capacitor.

It is interesting to quantify the power losses that these two networks introduce. They can be computed using (6.31) and (6.32) for the low pass and the high pass L-match networks, respectively.

$$\frac{P_{LOAD}}{P_{IN}} = \frac{(R_{OPT} - r_s)^2 + (\omega L_s)^2}{R_L R_{OPT}} \tag{6.31}$$

$$\frac{P_{LOAD}}{P_{IN}} = \frac{R_{OPT}}{R_L}\left[1 + \frac{1}{(\omega C_s R_{OPT})}\right] \tag{6.32}$$

L_S and C_S can be calculated from (6.33) and (6.34) for the low pass L-match network. Parameter r_S is the series parasitic resistor of the series inductor.

$$L_S = \frac{C_P R_L^2}{1 + (\omega_0 C_P R_L)^2} \tag{6.33}$$

$$C_P = \frac{1}{\omega_0 R_L}\sqrt{\frac{R_L - R_{OPT} + r_S}{R_{OPT} - r_S}} \tag{6.34}$$

L_P and C_S can be calculated from (6.35) and (6.36) for the high pass L-match network, where r_P is the parasitic resistor of the parallel inductor.

$$L_P = \frac{1}{\omega}\sqrt{\frac{(r_P + R_L)[(r_P + R_L)R_{OPT} - r_P R_L]}{R_L - R_{OPT}}} \tag{6.35}$$

$$C_S = \frac{(\omega_0 L_P)^2 + (r_P + R_L)^2}{\omega_0^2 L_P R_L^2} \tag{6.36}$$

Finally, the third topology, shown in Fig. 6.26, consists of a transformer that is made up of two integrated inductors. Expression (6.37) computes the network power losses with respect to the quality factors of the primary (Q_1) and secondary (Q_2) inductors and the coupling factor (k) [4].

$$\frac{P_{LOAD}}{P_{IN}} = \frac{1}{1 + 2\sqrt{\left[1 + \frac{1}{Q_1 Q_2 k^2}\right]\frac{1}{Q_1 Q_2 k^2}} + \frac{2}{Q_1 Q_2 k^2}} \tag{6.37}$$

Figure 6.27 compares the network losses of the previous three topologies for different values of the inductor quality factors. In order to establish a comparison, the inductors involved in each network have the same quality factor. The impedance transformation ratio is 3.33, downconverting 50–15 Ω at 5.5 GHz. For the integrated transformer, a coupling factor of 0.8 has been chosen [5].

As observed in Fig. 6.27, the transformer network losses are higher when compared to L-match networks. This is due to the relatively low value of the transformation ratio. In fact, when higher transformation ratios are necessary, a transformer network could be a good choice as its losses are quite independent of the transformation ratio, which is not the case for the L-match networks, where the losses increase with this ratio, as can be observed by comparing Fig. 6.27 with

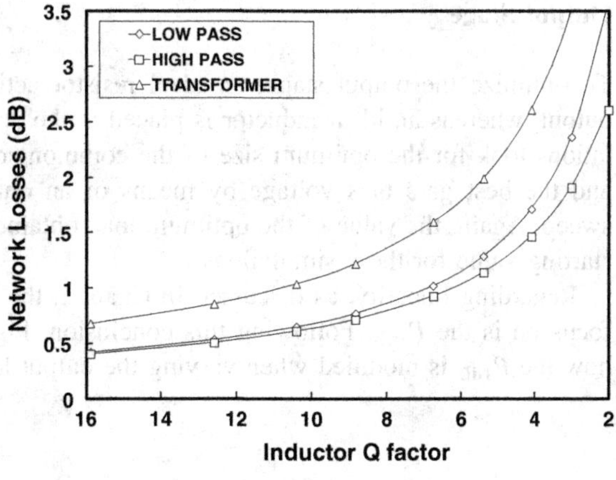

Fig. 6.27 High pass L-match, low pass L-match and transformer network losses against the involved inductor quality factors for a transformation ratio of 3.33

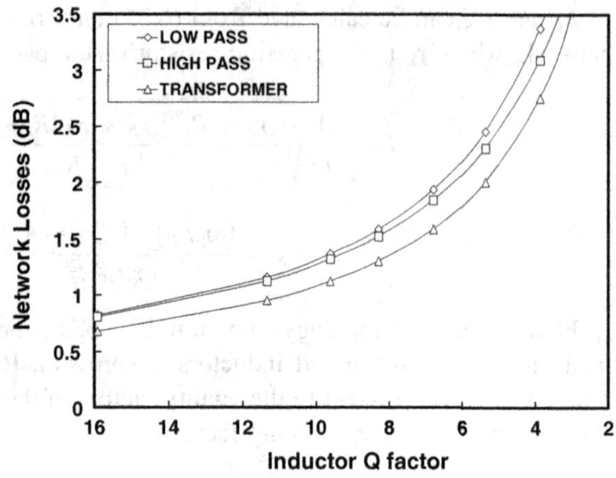

Fig. 6.28 High pass L-match, low pass L-match and transformer network losses against the involved inductor quality factors for a transformation ratio of 10

Fig. 6.28, where the transformation ratio has been increased to 10 (from 50 to 5 Ω).

With the help of the model, it is possible to choose the proper downconversion network. The model established a transformation ratio of 1.67, and therefore the use of any L-match network is a better choice than the transformer network. Specifically, for differential PAs the high pass L-match network presents the aforementioned advantage of size reduction when a balanced inductor is chosen when compared to the low pass L-match.

Moreover, it must be mentioned that the fact that the high pass L-match network shows slightly lower losses than the low pass L-match network cannot be regarded as an advantage because the differences are not significant and the low pass L-match network requires inductors with lower values, which usually translates into better qualities. Consequently, the losses of both networks will be similar in practice, with size reduction being the main advantage.

Output Stage

To optimize the output stage, an ideal resistor acting as R_{OPT} is placed at the output, whereas an ideal inductor is placed at the drain as a first step. The simulations look for the optimum size of the common-source and cascode transistors and the best gate bias voltage by means of an output load and drain inductor sweep. Again, the value of the optimum load obtained from the model is a useful starting value for these simulations.

Regarding linearity, as discussed in Chap. 2, the most important parameter to focus on is the P_{1dB}. Following this conclusion, Fig. 6.29 shows an example of how the P_{1dB} is modified when varying the output load of the output stage.

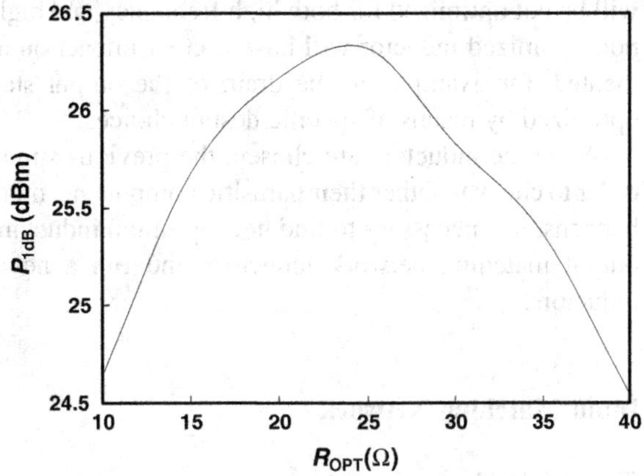

Fig. 6.29 Simulation results of the P_{1dB} parameter for different output load values at 5.25 GHz

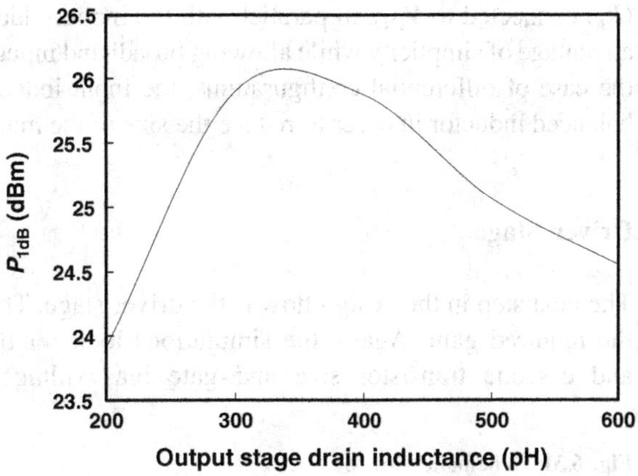

Fig. 6.30 Simulation results of the P_{1dB} parameter for different inductance values of the output stage drain inductor at 5.25 GHz

Figure 6.30 illustrates the influence of the output stage drain inductor in the P_{1dB} parameter. As can be seen, the P_{1dB} parameter is also sensitive to the drain inductance variations.

These simulations should be replicated for different transistor sizes and biasing values while keeping constant the current consumption to 300 mA, so that comparisons of the best size and biasing can be made.

Once the transistor size and the gate bias have been set, the next step is to simulate the required ideal values for the output parallel inductor and the series capacitors, following the high pass L-match network topology. When the ideal inductances for the drain and the output stage inductors have been found, it is necessary to replace them with real inductors. Although many CMOS processes offer a library of integrated inductors, these inductors are not always useful for PAs. The inductors within these libraries are usually implemented in the top metal layer with limited maximum widths and therefore may have electromigration issues for high current flows. Along with this, although they are wide enough to avoid electromigration, these inductors will be not optimized for both high frequency and high current flow applications. A non-optimized inductor will have a great impact on a PA if the inductor is implemented, for example, at the drain of the output stage. Therefore, they must be optimized by means of specific design choices.

When the inductors are chosen, the previous simulations should be repeated in order to check whether their parasitic components modify the optimal results. If this happens, it is necessary to find new optimum inductances for either the drain or the output matching network inductors and run a new simulation with these new inductors.

Input Matching Network

Figure 6.31 shows the components that comprise a useful input matching for full integration. The common-source transistors are biased by means of a low value resistor (R_g) connected to V_{DD}, in parallel with the input inductor (L_{IN}). This circuit has the advantage of simplicity while allowing broadband input matching. Along with this, in the case of differential configurations, the input inductor can be implemented as a balanced inductor in order to reduce the size of the matching circuitry.

Driver Stage

The next step in the design flow is the driver stage. This stage is designed to obtain the required gain. Again, the simulations look for the optimum common-source and cascode transistor size and gate bias voltage while sweeping the drain

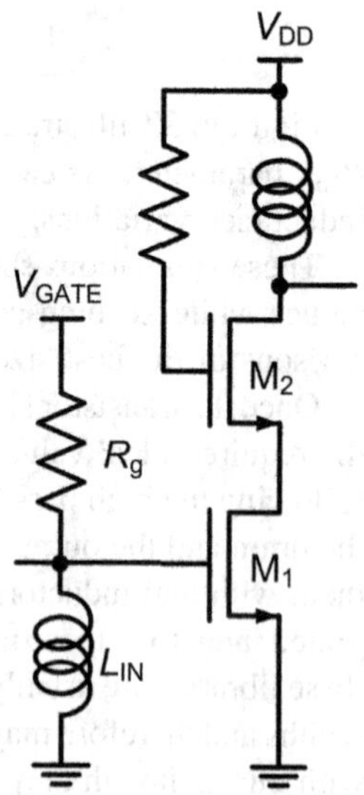

Fig. 6.31 Schematic view of the input matching for the single-ended equivalent circuit

inductor. As a first approximation, the output load of this stage is fixed at 50 Ω. The 50 Ω load will be obtained by means of interstage matching. This matching is required because the input load of the output stage is not large enough to permit high gain for the driver stage. In addition, the driver stage must not affect the linearity of the output stage. Regarding the PAE, the high gain and the relative low current consumption of this stage will not have a great impact on the overall PAE but the effect of the driver stage must be controlled so that the final PAE fulfills the specifications.

As explained before, the input matching of the PA is performed by means of a balanced inductor and the gate bias resistor dividers. The input matching inductor value is fixed when the common-source and cascode transistor sizes are known. As this inductor does not have to withstand high current levels, no extra consideration has to be taken into account.

Interstage Matching and Stability

The last step of the design flow is to combine the first stage with the output stage. At this step the stability of the PA must be also checked and ensured. In order to meet both needs, the interstage matching can be implemented by means of a decoupling capacitor and a series resistor. This series resistor avoids instabilities caused by the output capacitor of the driver stage along with its drain inductor, which sees the negative resistance provided by the output stage [6]. Even when the value of this resistor can be high, either the gain or the efficiency are only slightly affected because the resistor increases the output load of the driver stage and, therefore, its gain.

Finally, the cascode transistor gates are connected to the supply voltage but not directly. A series resistor is placed in series in order to stabilize the DC voltage at the gates of the cascode pair. If no series resistor were added, a voltage signal at RF would be coupled to the supply at the cascode transistor gates, affecting the PA performance. Along with this, the resistor also improves the stability of the PA. It must be also mentioned that as the PA can work in deep nonlinear regions, stability must be ensured for any power level and with the help of large signal S-parameter simulations because PA characteristics change for different power levels.

The next section describes the design flow of the inductors implemented in the PA, which have received the name of *power inductors*.

Power Inductor Design

As mentioned before, the design of a PA often requires the use of custom-designed integrated inductors because the inductor library offered by the foundry does not usually cover the requirements that a PA imposes. In fact, these components require some additional characteristics in order to be called power inductors.

These on-chip inductors must consider the following aspects:

1. If the power inductor is designed to be a drain inductor, it must present low DC resistance in order to minimize the supply voltage drop. As high DC currents flow through them, even a DC resistance increase of only one ohm drops the supply voltage seen by the transistors to several tenths of mV, which has a great impact on the efficiency and linearity characteristics of the PA, as has been observed in the model.
2. High Q values in the operating frequency band are essential for these kinds of inductors, since they are placed in parallel to the load. As deduced from the model, the high frequency resistance of these inductors affects the efficiency of the PA because they consume power that should be delivered to the load. Therefore, the quality factor curves must be optimized for the band of operation.
3. The power inductors must present large sections to the currents in order to avoid electromigration problems. That can be achieved by either increasing the width of the top metal layer or by increasing the thickness by adding new metal layers in parallel.

Top Metal Layer Width Considerations

Quite often the integrated inductors only make use of the thick top metal layer, commonly offered in modern RF CMOS processes, and are optimized by modifying only the geometrical characteristics of this metal track.

The advantage of using the thick top metal layer is that it withstands the highest current flows and the lowest resistivity, and a small width increase allows considerably higher current flows and a noticeable parasitic resistance reduction. However, the high current requirements of PAs can demand an extremely wide track that would degrade the inductor performance if only the last metal layer is used.

In addition to that, several studies [7, 8] recommend using metal track widths not exceeding 15–20 μm for high frequency applications, mainly due to the proximity effect, which degrades the inductor's Q performance. Even when this recommendation varies with each process characteristic—metal layer thickness, distance from the substrate, substrate conductivity, etc.—it is always true that there is an inductor track width beyond which only detrimental effects on the inductor's Q will be observed [8].

On the other hand, the process usually sets a width limit that should not be exceeded without the use of metal slots, which in turn decreases the actual width. A value of 20–30 μm is a typical width limit before metal slots are required to avoid metal stress issues. Furthermore, there is also an absolute maximum width that cannot be exceeded; 50 μm is a typical value.

Finally, Q enhancement techniques such as variable width [9] are far from practical in wide track power inductors. Therefore, the metal track width increase cannot be the only design choice in power inductors.

Multiple Metal Layer Considerations

In order to overcome the current flow constraint, while reducing the track width, the possibility of adding new metal layers in parallel must be also considered.

First, there are certain drawbacks that must be taken into account, e.g. the coupling capacitance and the skin effect increase. It is true that the use of several metal layers in parallel increases the coupling capacitance of the substrate; however, this effect is not high if only the last metal layers are used. Furthermore, as the increase in the number of stacked metal layers is followed by a track width reduction for the same current flow level, this effect is counteracted. Along with this, modern CMOS processes offer 6 or more metal layers, so that the last top metal layers are still far from the substrate. Several studies [7, 10] suggest moreover that is not worth using more than 2–3 metal layers because beyond this number the improvement is indiscernible. On the other hand, it is also recommended that some metal layers be kept for the underpass, so that no more than half of the total number of the available metal layers should be used.

Another drawback in the use of several metal layers is that the skin effect is more noticeable because the track width is reduced. However, the influence of the skin effect is negligible when compared to that of the proximity effect up to frequencies of 4–5 GHz [7]. Along with this, although the proximity effect between the stacked metal layers induces some losses, they are compensated by the fact that this effect decreases as the track width is reduced. In fact, some studies present inductor Q improvements when several metal layers are placed in parallel [10, 11].

Along with this, an interesting advantage of using several metal layers in parallel is that electromigration problems can be overcome with fewer metal stress problems. The reason is that metal stress is reduced when vias are used for metal layer interconnection, so that the process width limit for metal slots can be relaxed [12].

Finally, and especially for drain inductors, which require very low DC resistance the number of metal layers must be maximized. The reason comes from the fact that for a given inductance value and track section, the length of an inductor decreases as the number of stacked metal layers is increased. As a consequence, the value of the R_{DC} is reduced [13].

Therefore, for power inductors, and especially for those implemented as drain inductors, maximizing the number of metal layers is the best choice.

Inductor Geometry Considerations

Finally, there are some other helpful design rules regarding the geometry that must be considered in the optimization of the power inductor.

1. Inductor Internal Radius: Several studies state that integrated inductors should be designed to be hollow in order to reduce the influence of the proximity effect [14]. This rule is quantified in a value of 5 times the inductor width. However, it can be relaxed whenever the inductor does not exceed 2.5 turns [7]. For power inductors the possibility of relaxing this design rule is positive because it can lead to a reduction of the inductor area without affecting the inductor's Q performance.
2. Number of turns: The main effect caused by an increase in the number of turns is that the magnetic fields within the inductor are increased. Then it is possible to implement an inductor with the same inductance but less area for a specific inductance value. This area reduction means a smaller R_{DC} and a smaller oxide capacitance, so the Q curve can be shifted upward. This is usually required in order to set the maximum of the Q curve at a high frequency operation band. Therefore, the increase of the number of turns has a positive effect on power inductors.
   For small value inductors, an increase in the number of turns is not always possible, because the internal radius rule could not be fulfilled due to their relative small dimensions. However, this drawback is not crucial because the shifting up of the Q is not required due to the small dimensions and the R_{DC} is low enough due to the great track sections.
3. Number of sides and track spacing: The integrated inductor should be designed with the highest number of sides allowed by the process, because the inductor's performance improves in terms of inductance and Q [7]. At the same time, the spacing between tracks should be the smallest allowed by the process because it increases the inductance with no increase resistance. Although the lateral capacitive coupling between tracks must be also considered as it shifts down the Q curve, as a starting point the minimum spacing should be used.
4. Geometry of the vias: The geometry of the vias connecting in parallel the different metal layers also affects the Q performance of the inductor. The best configuration corresponds to longitudinal vias and, if the process does not allow them, discrete vias should be used along the whole metal track [7]. Both configurations are shown in Fig. 6.32.
   In addition, there is an extra advantage in the use of vias because, as has been mentioned before, it is possible to have an un-slotted track width beyond the one recommended by the process since irregularities caused by the vias on the track act in the same way as metal slots [12].
5. Ground shielding: Ground shielding has been proposed as an inductor Q enhancement technique, as it minimizes the losses of the magnetic field in the substrate by means of a plane made of metal, polysilicon or diffusion placed below the inductor. However, this technique is not always beneficial and can even degrade the Q of an unshielded inductor [11, 15, 16]. Furthermore, ground shielding has an important drawback since it increases the coupling capacitance

Fig. 6.32 Tri-dimensional view of longitudinal and discrete via configurations

to ground with a corresponding drop of the resonance frequency [16, 17]. In the case of inductors for PAs it is further aggravated by the fact that they require wide tracks and stacked metal layers for the turns and the underpass. Therefore, the use of ground shielding is not recommended for multi-layer inductors intended for PAs.

Some reports [18] show greater improvements in the Q with no resonance frequency drop. However, the results presented in [18] are valid only for differential inductors and require of at least one metal layer for the shielding.

Finally, the isolation from the substrate achieved by the ground shielding technique [16, 17] can be also solved by means of guard rings around the inductors without much affecting their performance.

The trade-off between the number of metal layers and the metal track width for a specific inductor, along with the amount of current it has to withstand is not an easy task. Therefore, although the use the previous rules help focus the optimal design, an electromagnetic simulator for these types of inductors is still necessary.

Accuracy Analysis of the Electromagnetic Simulator

The Agilent Momentum simulator is an appropriate option for performing planar inductor electromagnetic simulations. We checked the accuracy of Momentum by means of several simulations involving inductors fabricated in the same CMOS process. Two families of balanced and non-balanced inductors were used. The balanced inductors were labeled B1–B5, the non-balanced inductors were labeled NB1–NB5. Figure 6.33 shows a view of the chip that includes the balanced inductors.

The main parameters of the inductors appear in Table 6.2, where W indicates the inductor width, N the number of turns and S the spacing between turns. The internal and external diameters are also included. The inductors chosen cover

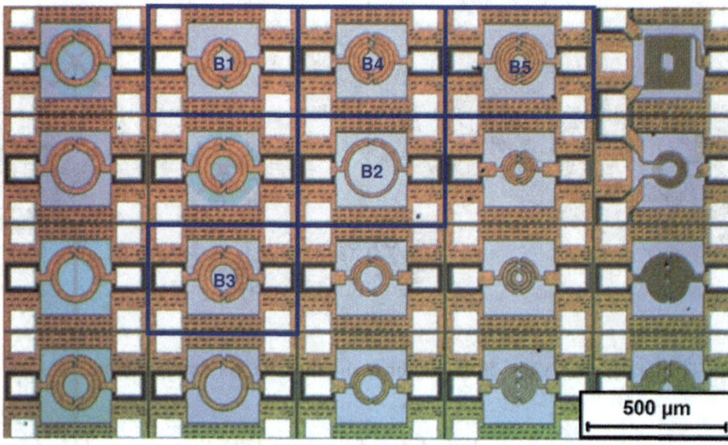

Fig. 6.33 Layout view of the balanced inductor library

Table 6.2 Description of the inductor set used for checking Momentum's accuracy

Ref.	M. Layer (Underpass)	W (µm)	InternalD (µm)	ExternalD (µm)	N	S (µm)
NB1	M6 (M5)	20	238	322	1.5	2
NB2	M6 (M5)	20	136	264	2.5	2
NB3	M6 (M5)	18	136	252	2.5	2
NB4	M6 (M5)	18	168	284	2.5	2
NB5	M6 (M5)	15	195	293	2.5	2
B1	M6 (M5)	20	74	200	2.5	1.5
B2	M6 (M5)	10	175	218	1.5	1.5
B3	M6 (M5)	20	94	220	2.5	1.5
B4	M6 (M5)	16	63	200	3.5	1.5
B5	M6 (M5)	14	48	200	4.5	1.5

inductance values from approximately 1 to 3 nH. This range of values was considered sufficient for the design of the power inductors suitable for a wide range of possible PA implementations.

Tables 6.3 and 6.4 show the results of two different simulations and the relative errors when compared to measurements. Table 6.3 contains the results and comparisons of a Momentum simulation using a closed boundary configuration for the substrate. Closed boundary means that the substrate has been configured in the Momentum process file with the bottom of the substrate connected to a GND perfect conductor. On the other hand, Table 6.4 shows the results and comparisons of a Momentum simulation using an open boundary configuration for the substrate. In this case, open boundary means that the substrate has an infinite thickness.

Table 6.3 Simulated results for the closed boundary configuration and relative errors with respect to measurements

Ref.	Simulated values						Relative percentage of error (%)					
	Inductance		Quality factor				Inductance		Quality factor			
	2.4 GHz	5 GHz	2.4 GHz	5 GHz	Q max	f@Max (GHz)	2.4 GHz	5 GHz	2.4 GHz	5 GHz	Q max	f@Max (GHz)
NB1	1.36	1.40	10.1	15.3	17.1	8.2	5.10	5.58	31.41	43.72	58.36	26.15
NB2	1.87	1.94	10.3	13.6	14.23	7.1	7.66	6.24	27.27	36.42	41.86	44.90
NB3	1.86	1.92	9.8	13.4	14.2	7.2	7.02	5.73	23.41	32.33	40.04	46.94
NB4	2.28	2.40	10.1	12.9	12.9	5.6	6.19	4.03	18.99	33.49	30.58	43.59
NB5	2.68	2.87	9.5	11.9	11.9	5.1	5.51	3.68	12.67	28.75	23.39	30.77
B1	0.97	0.96	6.5	10.5	16.2	>10	4.90	6.80	9.72	5.00	7.28	–
B2	1.02	1.02	6.3	11.4	16.8	>10	4.67	5.56	3.08	6.56	26.96	–
B3	1.17	1.17	7.3	11.9	17.5	>10	7.14	9.30	14.12	0.85	15.89	–
B4	1.61	1.63	6.3	9.2	11.4	9.6	9.04	9.44	16.33	11.54	4.20	–
B5	2.24	2.33	6.1	7.9	8.4	7.2	4.27	9.34	4.69	0.89	3.45	26.32

Table 6.4 Simulated results for the open boundary configuration and relative errors with respect to measurements

	Simulated values						Relative percentage of error (%)					
Ref.	Inductance		Quality factor				Inductance		Quality factor			
	2.4 GHz	5 GHz	2.4 GHz	5 GHz	Q max	f@Max (GHz)	2.4 GHz	5 GHz	2.4 GHz	5 GHz	Q max	f@Max (GHz)
NB1	1.46	1.49	9.8	13.0	13.3	6.2	12.83	12.37	27.50	22.11	23.17	4.62
NB2	1.96	2.03	10.2	12.5	12.6	5.8	12.84	11.17	26.03	25.39	25.87	18.37
NB3	1.93	2.00	9.8	12.5	12.7	6.2	11.05	10.13	23.41	23.44	25.25	26.53
NB4	2.37	2.51	9.80	12.20	12.20	5.3	10.39	8.80	15.46	26.24	23.49	35.90
NB5	2.77	2.95	9.5	11.5	11.5	5.0	9.06	6.58	12.67	24.42	19.25	28.21
B1	0.96	0.95	6.2	8.9	10.7	9.7	5.88	7.77	13.89	11.00	29.14	–
B2	1.01	1.02	6.2	10.8	14.9	>10	5.61	5.56	4.62	11.48	35.22	–
B3	1.24	1.24	7.3	10.9	13.2	9.8	1.59	3.88	14.12	7.63	12.58	15.29
B4	1.67	1.69	6.3	8.9	10.3	8.9	5.65	6.11	16.33	14.42	13.45	–
B5	2.30	2.39	6.1	7.8	8.1	6.9	1.71	7.00	4.69	0.38	0.25	21.05

Tables 6.3 and 6.4 show that inductance errors are low, namely below 10 % for the typical closed boundary configuration. Conversely, the errors for quality factor and frequency of maximum Q values are significantly higher. It must be mentioned that moderate Q errors are not so crucial, first because an error of 2 or 3 units for a measured Q value of 10 can be accepted, which means a 20–30 % error rate. Second, because Q curves are usually quite flat for a wide frequency range, a maximum frequency error of, approximately, 1 GHz in the 5 GHz band can be allowed, which means 20 % in frequency error.

However, it is possible to improve the accuracy of the simulations if the substrate shows infinite thickness. If the open boundary configuration is used, the simulations introduce extra losses and, therefore, the quality factor values and the relative errors decrease. This accuracy improvement also allows the Q curve to be centered more exactly on the frequency band of interest. Along with this, inductance errors for open boundary configuration are still low if Tables 6.3 and 6.4 are compared.

Results from Tables 6.3 and 6.4 indicate that Momentum is a valid electromagnetic simulator, because the inductance errors stay at around 10 % for inductance and quality factor measurements if the open boundary configuration for the substrate is used.

Finally, as Table 6.2 illustrates, only top metal layer (M6) has been used for inductor simulations and measurements, but power inductors could require the use of several metal layers in parallel. However, it is expected that Momentum will show similar accuracy as that presented for the previous inductors. On the other hand, as observed in Tables 6.3 and 6.4, some values regarding the frequency of maximum Q do not appear. The reason is that this frequency is beyond 10 GHz in measurements or in simulations.

Power Inductor Design Flow

Power inductors are implemented as drain inductors for the driver and the output stages of the PA and for the output matching network. The design flow of these inductors is as follows:

1. First, an ideal inductor value is placed in the schematic view of the PA and an optimal ideal inductance value is obtained from simulations. As a first step, it is also essential to keep the maximum DC (I_{DC}) and peak RF current ($I_{RF,pk}$) flowing through them, as it will set the width and number of metal layers of the inductor.
2. The next step is to determine the number of metal layers and the width that the inductor shall have. This decision is based on the aforementioned rules along with the values of the DC and RF peak current levels and the current density of the metal layers in order to avoid electromigration issues.

3. Any integrated process establishes a maximum current density (J_{MAX}) that can flow through a metal layer in order to avoid electromigration issues. Current density parameter is given in mA/µm units for each metal layer. The minimum width (W) of a specific metal layer can be calculated by (6.38).

$$W(\mu m) = 1.2\frac{I_{DC} + \frac{I_{RF,pk}}{\sqrt{2}}}{J_{MAX}(mA/\mu m)} \tag{6.38}$$

Where the 1.2 factor is included as an extra margin.

If the metal layers are placed in parallel, their current density values are added for width calculations, so that the more layers there are in parallel, the smaller the final width of the inductor.

In the Momentum simulations, vias are simplified using the external walls so the edge of a metal layer is connected to the adjacent one, as illustrated in Fig. 6.34. It can differ from typical CMOS process implementations, in which vias are shaped as multiple squares. However, multiple square vias would slow down Momentum simulations excessively, and this simplification facilitates the simulations while achieving accurate results.

As an implementation example, Fig. 6.35 shows within the test fixture the layout of a power inductor intended for the drain of a high current driver stage of a PA. Therefore, the design of this inductor aims for both high Q at the operating frequency and low DC resistance.

This inductor must withstand 160 mA of DC current, so in order to avoid electromigration problems, the last three metal layers (M6, M5 and M4) have been used in parallel to implement the inductor that has 1.5 turns and is 28 µm wide. The other three metal layers (M3, M2 and M1) have been used for the underpass. Note that the underpass has a different width. The reason is that the overall maximum J_{MAX} of the top three metals layers is higher than that of the first three ones.

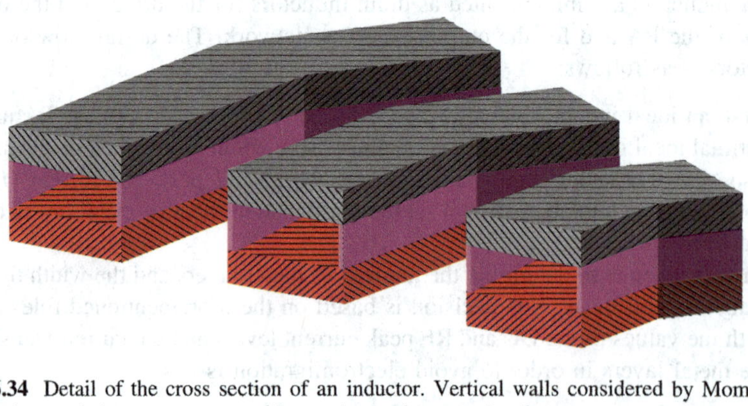

Fig. 6.34 Detail of the cross section of an inductor. Vertical walls considered by Momentum appear in *purple* between two metal layers (*striped grey* and *red*)

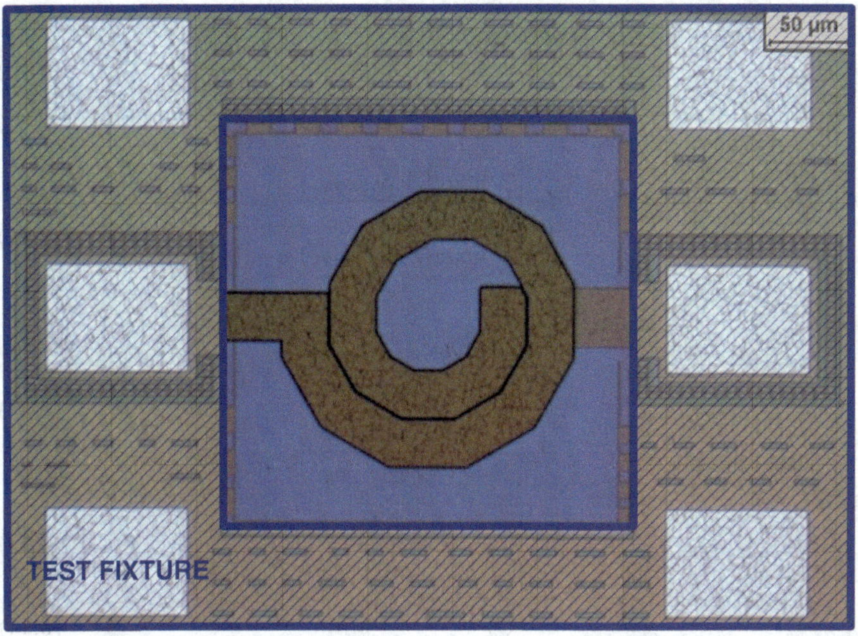

Fig. 6.35 Non-balanced power inductor layout with a width of 28 μm, three metal layers and 1.5 turns inside its test fixture

It must also be noted that, as mentioned before, although the inductor width exceeds the limit beyond which metal slots are required, it is not problematic because the use several metal layers connected with vias reduces the metal stress [12].

Finally, following the previous rules, in order to reduce the inductor area, the internal radius is less than two times the track width with no Q degradation and uses the maximum number of sides and the minimum track spacing allowed by the integrated process.

Momentum requires the thickness, conductivity and permittivity of the process dielectric and metal layers to perform the simulations. A process file that included the available metal layers, $Si0_2$ dielectric layers and lossy substrate was implemented. By means of this file and the layout implementation, Momentum gives the S-parameters as a result of the simulations. Those parameters can be translated into quality factor (Q) and inductance (L) values by means of Eqs. (6.39–6.44).

$$\begin{aligned} S_{1,ONEPORT} &= S_{11} - \frac{S_{12}S_{21}}{S_{22}} \\ S_{2,ONEPORT} &= S_{22} - \frac{S_{12}S_{21}}{S_{11}} \end{aligned} \tag{6.39}$$

$$Z_{1,ONEPORT} = 50\frac{1 + S_{1,ONEPORT}}{1 - S_{1,ONEPORT}}$$
$$Z_{2,ONEPORT} = 50\frac{1 + S_{2,ONEPORT}}{1 - S_{2,ONEPORT}} \tag{6.40}$$

$$L_{11} = \frac{\mathrm{I}mag(Z_{1,ONEPORT})}{2\pi f}$$
$$L_{22} = \frac{\mathrm{I}mag(Z_{2,ONEPORT})}{2\pi f} \tag{6.41}$$

$$Q_{11} = \frac{\mathrm{I}mag(Z_{1,ONEPORT})}{\mathrm{R}eal(Z_{1,ONEPORT})}$$
$$Q_{22} = \frac{\mathrm{I}mag(Z_{2,ONEPORT})}{\mathrm{R}eal(Z_{2,ONEPORT})} \tag{6.42}$$

$$L_T = \frac{L_{11} + L_{22}}{2} \tag{6.43}$$

$$Q_T = \frac{Q_{11} + Q_{22}}{2} \tag{6.44}$$

As (6.39) to (6.44) show, inductance and quality factor values are a mean of the values measured from both ports of the inductor. That is advisable for non-balanced inductors in which the L and Q curves are different when measured from one port or the other.

It is possible to implement the model of the simulated inductor in CADENCE by means of the S-parameters. However, CADENCE cannot perform nonlinear simulations with the S-parameter model so it is necessary to obtain a lumped model for the inductors. In order to obtain a lumped model, the compatibility between Momentum and ADS is used with an ADS schematic view in which a generic PI-model is compared to the S-parameter file from Momentum. The ADS optimization process equates the performance of the PI-model to that of the Momentum S-parameter file by sweeping the values of the PI-model's lumped components. The model presented in Fig. 6.36 shows the components of the PI-model used in the ADS fitting process. Finally, the lumped PI-model is copied to CADENCE.

Finally, the description and measured results of some power inductors will be presented and compared to measurements in the following chapter.

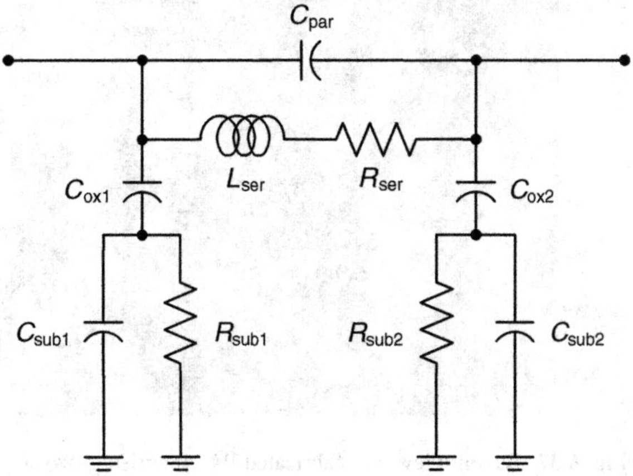

Fig. 6.36 Schematic view of the PI-model utilized for the inductors. C_{par} refers to the coupling capacitance between windings. L_{ser} is the inductance and R_{ser} the parasitic resistance. C_{ox} refers to parasitic capacitance to the substrate and C_{sub} and R_{sub} models the substrate

Layout Design

The layout of a PA presents several issues that must be taken into account to avoid performance degradation from the expected results.

Circuit Isolation

The overall layout of the PA must be surrounded by a guard ring implemented with all the available metal layers connected to ground and to the substrate. The purpose of this ring is to isolate the PA from the other circuits presented in the same die and the ring should be kept when connected to the other circuits of the transceiver. In Fig. 6.37 an example of a PA layout is shown. As can be observed, the ring is not closed in order to keep current from flowing through a low impedance closed loop, which would create an electromagnetic field and affect the performance of the PA.

Differential Design Considerations

As was previously mentioned, a differential design is an appropriate choice for the implementation of a linear PA. In order to keep the differential characteristics of the circuit some rules must be followed.

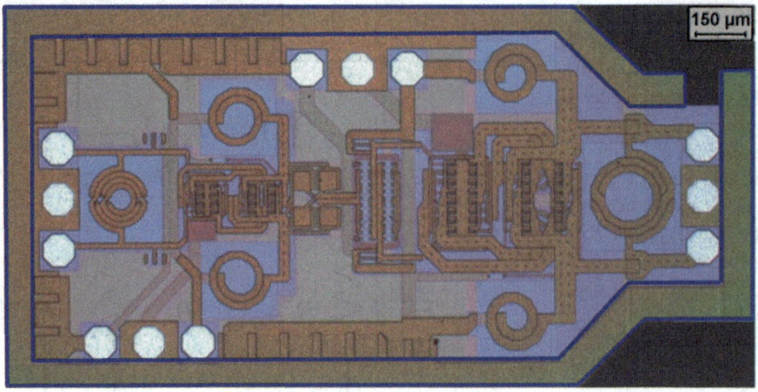

Fig. 6.37 Layout view of a fabricated PA comprising two stages, the input and output matching networks and the interstage network. The PA is surrounded by a wide multi-layer metal track connected to ground

Balanced Inductors

As Fig. 6.38 shows, in order to keep the topology of the PA differential the input matching network should be composed of a balanced inductor. The use of a balanced inductor also serves to save extra area in differential architectures.

The balanced inductor of the output stage is shown in Fig. 6.39. As can be observed, the series capacitors of the impedance downconversion network are placed above and below the balanced output inductor, and they are placed as close as possible to the inductor in order to reduce any parasitic inductance. The metal track interconnection has been implemented using the top metal (M6), because it presents the greatest thickness. It allows a smaller width for the tracks, which is important due to the high current levels that flow through them. However, as it was necessary to implement tracks wider than 20 µm, slots were required in order to avoid stress problems during the fabrication process [12].

Common-Centroid

A differential topology is improved if a symmetric configuration is implemented, as is the case of the balanced inductors in both the driver and the output stages. But a differential topology is not ensured only by means of a symmetric configuration because process dispersions do not act everywhere in the chip equally. A common-centroid configuration for all the components placed in the differential branches should also be implemented. Figure 6.40 illustrates an example of this topology for the common-source and cascode transistor of a PA.

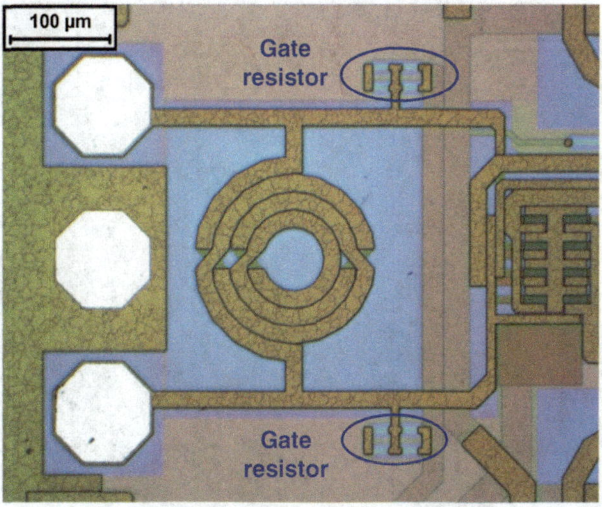

Fig. 6.38 Layout of view of the input matching network, which comprises a balanced top metal layer inductor and the gate resistor differential pair

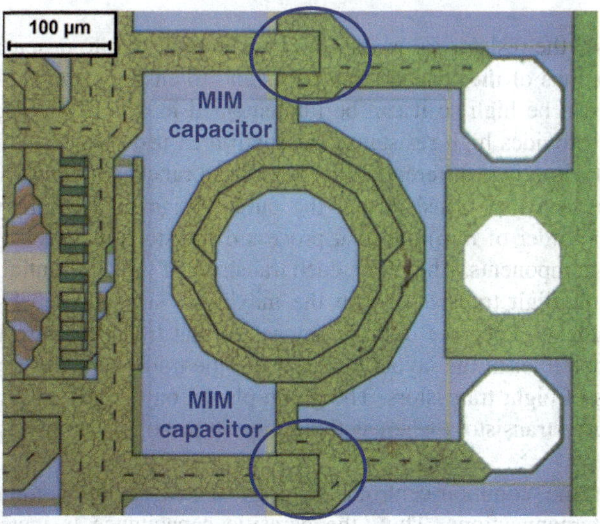

Fig. 6.39 Layout of the output matching network, which comprises a balanced three metal layer inductor and the MIM capacitor differential pair

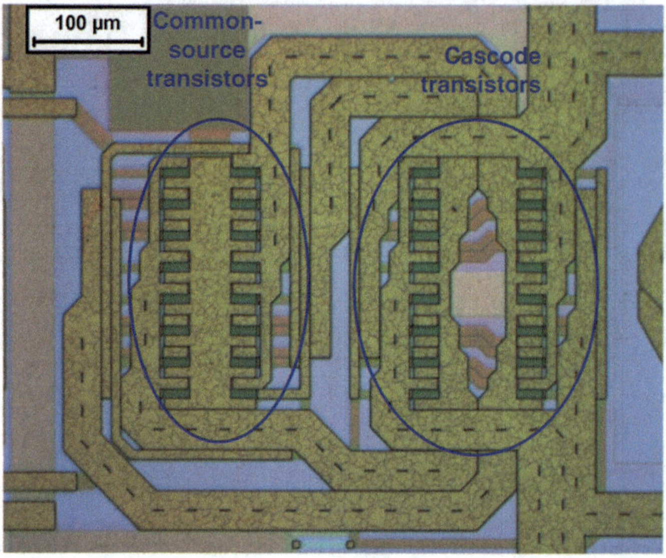

Fig. 6.40 Layout of the output stage based on the common-source and cascode transistor pairs. Interconnection between cascode transistors and supply voltage through the RC resistor is also shown

In Fig. 6.40 the resistor R_C is also shown. It serves both to desensitize the gate voltage variations of the cascode transistors and to ensure stability. The value of this resistor can be high so it can be implemented in a non-silicided polysilicon layer, which provides high resistance for small area resistors.

In order to improve differentiability between branches, it must also be taken into consideration that transistors in the output stage of a PA may have great widths (in the order of mm) and that process dispersions increase with the separation of the components. Therefore, each transistor of the differential pair must be formed by interdigit transistors with the maximum size allowed by the CMOS process in order to keep the design compact so that transistors are kept as close together as possible. In the layout of Fig. 6.40, the transistors are implemented in two columns of eight transistors. The group placed on the left corresponds to the common-source transistors whereas the group on the right comprises the cascode ones.

Note that the common-centroid topology increases the complexity of the transistor interconnections. Thus, the parasitic capacitance is increased, as the topology requires lower metal layers be used and the crossing of the different metal tracks be increased. However, this topology ensures that the differential performance is maintained despite process dispersions. The specific common-centroid implementation reduces the parasitic capacitance by minimizing both the metal track crossings and the metal track width.

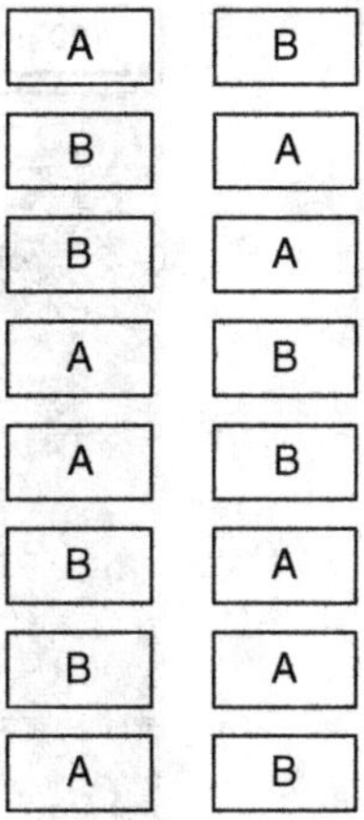

Fig. 6.41 Transistor distribution of the common-centroid configuration used for both the common-source and the cascode transistors of the output stage

With regard to the metal tracks, in order to minimize parasitic capacitance, they must be dimensioned just to withstand the current flowing through them. Thus, when a metal track diverts a certain amount of current to, for example, a transistor drain, the track width is reduced.

Again, the use of slots in the metal tracks is required by the high current levels, with the slots being placed in parallel to the current flow in order to minimize the impact on the current density.

Finally, a distribution of the common-source and cascode transistors used in the common-centroid configuration is illustrated in Fig. 6.41. Either the common source or the cascode transistor pairs follow the same transistor distribution. The transistors named A belong to one of the differential branches whereas the B transistors belong to the other.

Passive Components

Passive components implemented in high frequency differential PAs will have specific characteristics. In the following sections, resistors and capacitors are discussed.

Low Value Resistors

As mentioned before, low value resistors are required at least for the driver and the interstage matching networks. These resistors are usually based on the polysilicon layer of the CMOS transistor gates but with the silicide implant, that is used to reduce the resistivity, blocked. Therefore, resistivities of about 100 Ω/square can be achieved. When these resistors are going to be used in differential topologies it

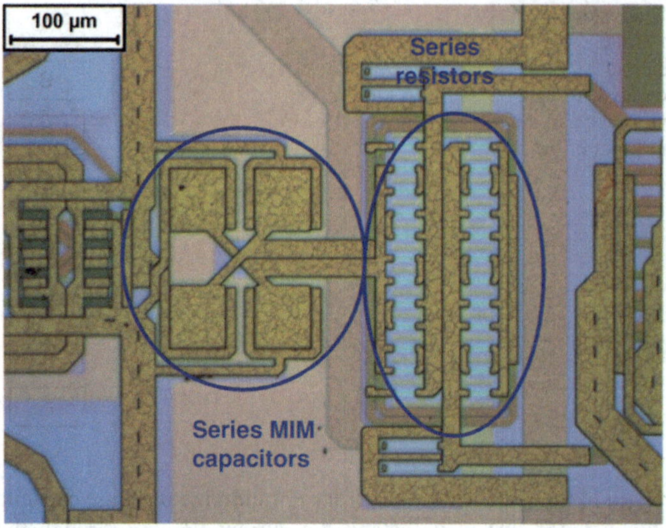

Fig. 6.42 Layout of the interstage matching, which comprises the series capacitor and series resistor pairs in common-centroid configuration

is essential that the effects of process dispersions be reduced as much as possible. In order to reduce these effects, which affect differentiability, a length/width factor of 3–5 and a width five times the process minimum feature size rules should be followed [12]. Thus, sometimes it is necessary to put several resistors in parallel in order to obtain low values with these kinds of resistors. Another way to reduce resistor tolerances during the fabrication process is the use of dummy polys at both sides of the resistors, which is the case of all resistors used in this design [12].

An example of the combination of common-centroid configuration, the use of several resistors in parallel and dummy polys is illustrated in Fig. 6.42 for the interstage matching resistor. In this case, twelve non-silicide polysilicon resistors in parallel for each differential branch have been used. The distribution of the common-centroid configuration is illustrated in Fig. 6.43.

RF Capacitors

The capacitors that are used for the output and the interstage network implementation are Metal–Insulator-Metal (MIM) capacitors. They are composed of two parallel plates using one of the last metal layers and an auxiliary metal layer. This extra metal layer is extremely close to the other so that the capacitance per area is relatively high ($1\text{fF}/\mu\text{m}^2$), making the implementation of several pF capacitors possible.

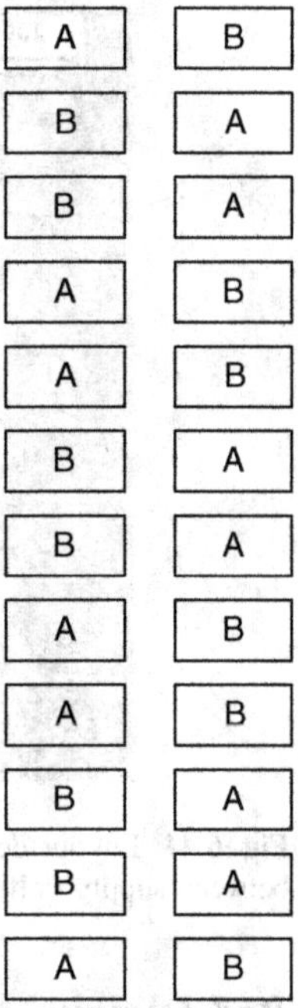

Fig. 6.43 Series resistor pair distribution of the common-centroid configuration

Some processes also offer MOS capacitors derived from high area MOS transistors. Their advantage is that they present higher capacitance per area. However, they have worse quality factors, as their plates show more series resistance; furthermore they are closer to the substrate increasing parasitic capacitance. Consequently, they are not suitable for high frequency applications and usually there is no RF model for them.

Stage Isolation

As mentioned before, the implemented series resistor between stages improves stability. However, the stability of a PA can decrease if the amplifier stages are not well isolated, since signals could travel through the substrate, creating a feedback loop that decreases isolation and consequently the stability factor K. In order to avoid that, physical separation and substrate contacts between stages must be implemented, as illustrated in Fig. 6.44.

Finally, high value coupling capacitors should be placed between ground pads and the different supply voltage pads as illustrated in Fig. 6.44.

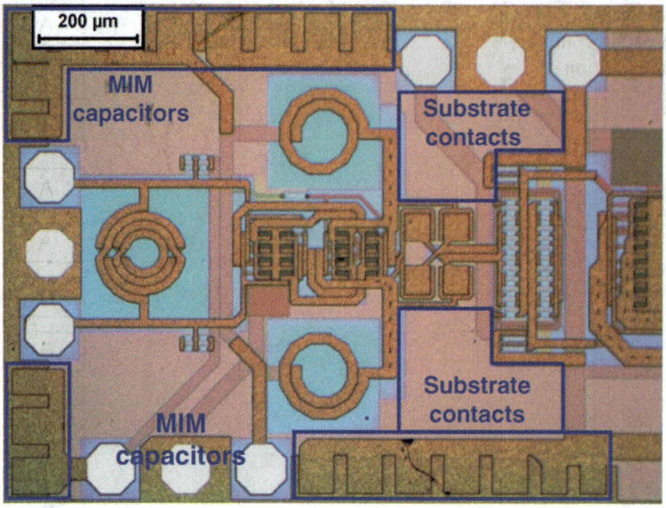

Fig. 6.44 Layout detail of the substrate contacts between stages and the coupling capacitors between supply voltage pads and ground

Pad Design

The pads used for circuit characterization are not usually implemented in the final transceiver design, in which ESD protection is also required. However, in order to avoid a loss of performance in the testing process, some considerations must be taken into account. First, it is recommended that the pads be designed as octagonal shapes, which reduces their parasitic capacitance without affecting the required testing area. There are three different configurations for the three different possible types of pads: GND (Ground) pads, supply and bias voltage pads, and RF pads. Their cross sections are illustrated in Fig. 6.45.

The pads designed for GND present all the six metal layers connected to the substrate by contacts.

The pads used for DC signals use the six metal layers as well in order to increase their robustness. Furthermore, the relatively high parasitic capacitance of these pads to ground is useful, as it helps stabilize the different supply and gate bias voltages.

The RF pads comprise the last two metal layers (M6 and M5) in order to make them robust enough for the testing process. No more metal layers are necessary as the top metal (M6) presents great thickness and minimum parasitic capacitance is desired. The first metal layer (M1) is placed below the RF pad in order to isolate it from the substrate. It is normally connected to ground, but if this connection is

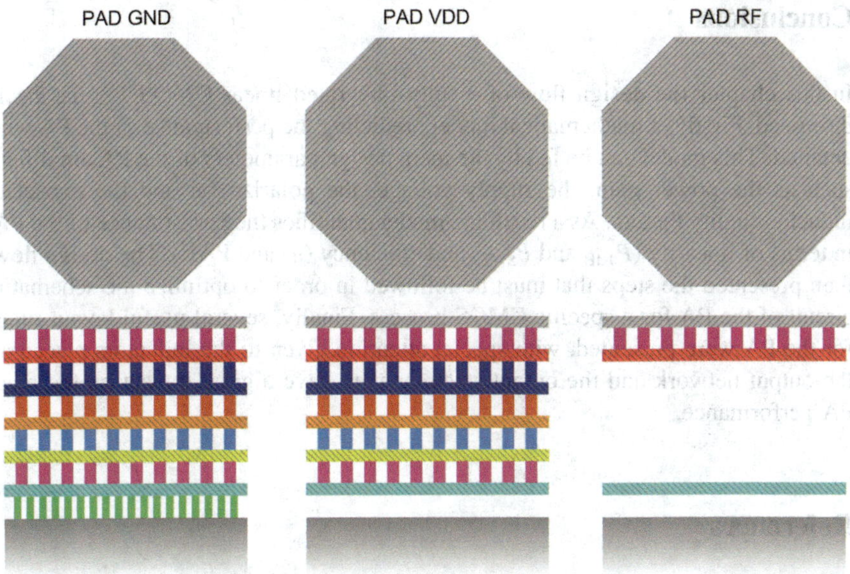

Fig. 6.45 Cross section of the different pads used in the design

avoided the parasitic capacitance can be reduced because metal-to-metal and metal-to-substrate parasitic capacitances are now placed in series.

A comparison of the V_{DD} and RF pad models is shown in Fig. 6.46. As can be observed, the parasitic capacitance is greatly reduced in the case of the RF pad.

Fig. 6.46 Schematic models of the V_{DD} and RF pads. The models are based on a parasitic capacitor to the substrate in series to a substrate resistance to ground

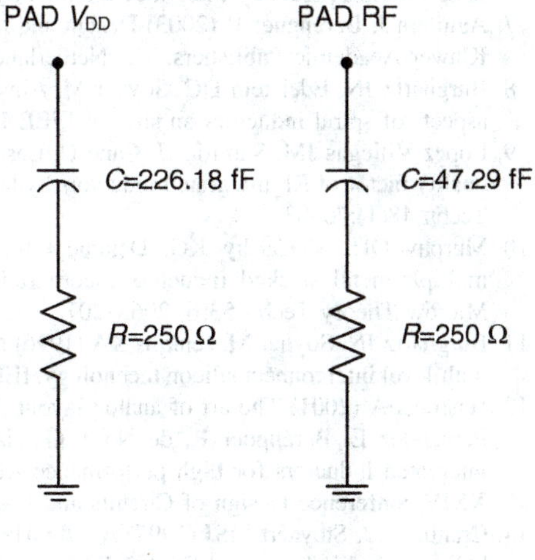

Conclusions

In this chapter the design flow of a fully integrated linear CMOS PA has been discussed. Firstly, a mathematical model predicting the performance of the PA was detailed. This model can be fed by the main design parameters of the PA amplifier, such as the power gain, the supply voltage, the polarization and the expected inductor quality factors. As a result the model quantifies the performance of the PA in terms of linearity (P_{1dB} and P_{SAT}) and efficiency (η_d and PAE). The design flow then presented the steps that must be followed in order to optimize the schematic circuit of the PA for a specific CMOS process. Finally, several useful layout rules for the PA were presented, with special attention given to the inductors placed at the output network and the output stage, as they have a great impact on the final PA performance.

References

1. Mazzanti A, Larcher L, Brama R, Svelto F (2006) Analysis of reliability and power efficiency in cascode class-E PAs. IEEE J Solid-State Circuits 41(5):1222–1229
2. Cripps SC (1999) RF power amplifiers for wireless communications. Artech House, Norwood
3. Reynaert P, Steyaert M (2006) RF power amplifiers for mobile communications. Springer, Dordrecht
4. Aoki I, Kee SD, Rutledge DB, Hajimiri A (2002) Distributed active transformer—a new power-combining and impedance-transformation technique. IEEE Trans Microw Theory Techn 50(1):316–331
5. Simburger W, Heinz A, Wohlmuth HD, Bock J, Aufinger K, Rest M (2000) A monolithic 2.5V, 1W silicon bipolar power amplifier with 55% PAE at 1.9 GHz. In: IEEE MTT-S International Microwave Symposium Digest (IMS 2000), pp 853–856
6. Adabi E, Heydari B, Bohsali M, Niknejad AM (2007) 30 GHz CMOS low noise amplifier. In: IEEE Radio Frequency Integrated Circuits Symposium (RFIC 2007), pp 625–628
7. Aguilera J, Berenguer R (2003) Design and test of integrated inductors for RF applications. Kluwer Academic Publishers, The Netherlands
8. Burghartz JN, Edelstein DC, Soyuer M, Ainspan HA, Jenkins KA (1998) RF circuit design aspects of spiral inductors on silicon. IEEE J Solid-State Circuits 33(12):2028–2034
9. Lopez-Villegas JM, Samitier J, Cane C, Losantos P, Bausells J (2000) Improvement of the quality factor of RF integrated inductors by layout optimization. IEEE Trans Microw Theory Techn 48(1):76–83
10. Murphy OH, McCarthy KG, Delabie CJP, Murphy AC, Murphy PJ (2005) Design of multiple-metal stacked inductors incorporating an extended physical model. IEEE Trans Microw Theory Techn 53(6):2063–2072
11. Burghartz JN, Soyuer M, Jenkins KA (1996) Microwave inductors and capacitors in standard multilevel interconnect silicon technology. IEEE Trans Microw Theory Techn 44(1):100–104
12. Hastings A (2001) The art of analog layout. Prentice Hall, Upper Saddle River
13. Fernandez E, Berenguer R, de No J, Garcia J, Solar H (2009) Design considerations of integrated inductors for high performance RF CMOS power amplifiers. In: Proceedings of XXIV conference Design of Circuits and Integrated Systems (DCIS 2009), pp 203–208
14. Craninckx J, Steyaert MSJ (1997) A 1.8-GHz low-phase-noise CMOS VCO using optimized hollow spiral inductors. IEEE J Solid-State Circuits 32(5):736–744

15. Chen YE, Bien D, Heo D, Laskar J (2001) *Q*-enhancement of spiral inductor with N+-diffusion patterned ground shields. In: IEEE MTT-S International Microwave Symposium Digest (IMS 2001), vol 1282, pp 1289–1292
16. Burghartz JN (1998) Progress in RF inductors on silicon-understanding substrate losses. In: IEEE International Electron Devices Meeting (IEDM 1998), pp 523–526
17. Yue CP, Wong SS (1998) On-chip spiral inductors with patterned ground shields for Si-based RF ICs. IEEE J Solid-State Circuits 33(5):743–752
18. Cheung TSD, Long JR (2006) Shielded passive devices for silicon-based monolithic microwave and millimeter-wave integrated circuits. IEEE J Solid-State Circuits 41(5):1183–1200

Chapter 7
Test Setups and Results

Abstract This chapter presents and discusses the measurements that are required to characterize PAs and power inductors. The setups required to perform the tests are also described, along with several recommendations for improving the accuracy of integrated circuit measurements.

Equipment

The main RF equipment required to perform the tests that characterize the inductors and the PA is listed in Table 7.1. Additionally, Agilent Advanced Design System (ADS) and Vector Signal Analysis (VSA) 89600 software tools are used to measure a PA with a digital channel. The following sections describe the test setup, the procedure for characterizing the circuit and the results.

Inductor Characterization

Accurately characterizing the inductors implemented in an integrated PA is extremely important in order to correctly evaluate the final performance of the PA in terms of output power, efficiency and linearity. The characterization of such inductors requires a careful procedure in order to obtain accurate measurements. The parasitics of the test fixture in which the inductor is embedded must be completely removed. When the inductors are intended for a PA with wide metal tracks and several stacked metal layers, correct test procedure is crucial; otherwise the errors made during the characterization process will impact on the design of the PA, introducing large errors between simulated results and measurements. In the following sections, the correct setup for inductor characterization is detailed, and some useful recommendations for getting accurate results are given.

H. Solar Ruiz and R. Berenguer Pérez, *Linear CMOS RF Power Amplifiers*, DOI: 10.1007/978-1-4614-8657-2_7,

Table 7.1 Description of the equipment used in taking PA measurements

MODEL	DESCRIPTION
Agilent E4438C	Vector signal generator, 250 kHz to 6 GHz
Agilent E4440A	Spectrum analyzer, 3 Hz to 26.5 GHz
Agilent E5071A	Vector network analyzer, 300 kHz to 8.5 GHz
Agilent E8408A	VXI mainframe
Agilent E8491B	IEEE 1394 port module
Agilent 89605B	Input module
Agilent E1439A	95 MSa/s ADC + FILTER + FIFO + 70 MHz IF input module
Agilent E3646A	DC power supply
Keithley 2000	Digit multimeter
Anaren 30057	4–8 GHz 180° hybrid coupler
Weinschel 7006	100 kHz to 20 GHz DC block
Weinschel 41-30-12	DC to 18.0 GHz 1 W 30 dB attenuator
Cascade SUMMIT 9101	Probe station
Cascade ACP40-D	150 μm SGS-type microprobe
Agilent 82357A	GPIB/USB adapter

Test Setup

Figure 7.1 shows the basic setup for characterizing an integrated inductor at RF frequencies. The test equipment consists of a Vector Network Analyzer (VNA), a probe station and the RF probes.

The VNA allows the S-parameters of the inductor to be measured, which will be further processed in order to obtain the inductor's inductance and quality factor.

The probe station ensures that the RF probes can be placed onto the inductor with highly accurate alignment, and it tightly controls the pressure applied to the probes as well. Accurate alignment and correct pressure are crucial for accurate results. In manual probe stations, alignment and pressure are enabled by means of the x, y and z micrometers. Along with this, the planarity micrometer allows the planarity of the probes to be adjusted.

The probes acts as an adapter between the traditional coaxial interface used with RF cables and the contact pads placed on the chip on-wafer. For RF measurements the most common probe is the air coplanar probe (ACP). As the name indicates, ACPs are implemented as a coplanar waveguide in air. When the pads are made of aluminum, the tips of the probe are usually made of tungsten (W). Because the tungsten (W) probes are firm, they break through the oxide film and make good electrical contact with the pad [1].

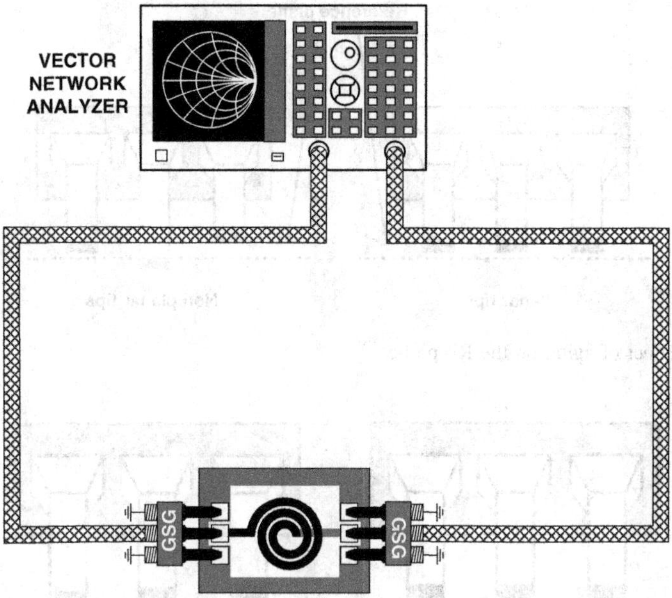

Fig. 7.1 Setup for inductor characterization

Prior Steps

The procedure for accurately characterizing inductors starts with considering two aspects related to the probe, namely planarization of the probe and skating.

Planarization

Before proceeding with any measurement, the first step is to check the planarity of the probes. One of the reasons is the probes may be aged, which often means that the tips are often no longer on the same plane. Figure 7.2 shows the case of a Ground-Signal-Ground (GSG) probe that has worn out. As we can see, the reference plane at the probe tips is no longer a straight line. This poor reference plane definition will introduce large errors in the inductor's measurement [1].

The common way to check probe tip planarity is by means of a contact substrate [2]. The contact substrate is simply a metalized field, and when the probes are lowered onto this field, the probe tips leave scratches in the metal, as shown in Fig. 7.3. Different scratch depths reveals that the probe tips are not on the same plane. If the scratches show that the lack of planarization is not caused by aging but rather it is only because the probe is rotated with respect to chip plane, the planarity micrometer allows the effect to be corrected, as is shown on the left side of Fig. 7.3.

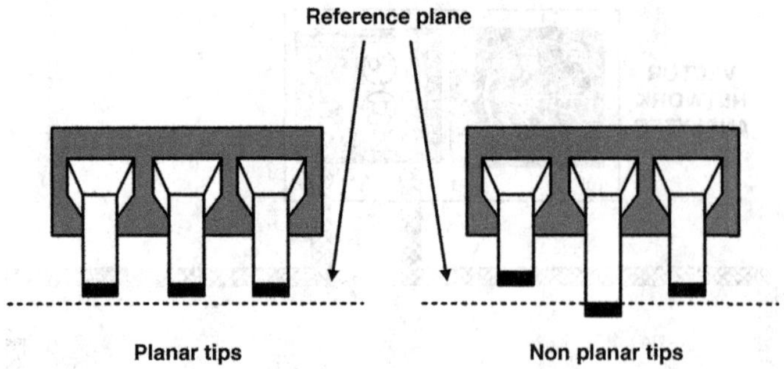

Fig. 7.2 Effect of aging on the RF probe

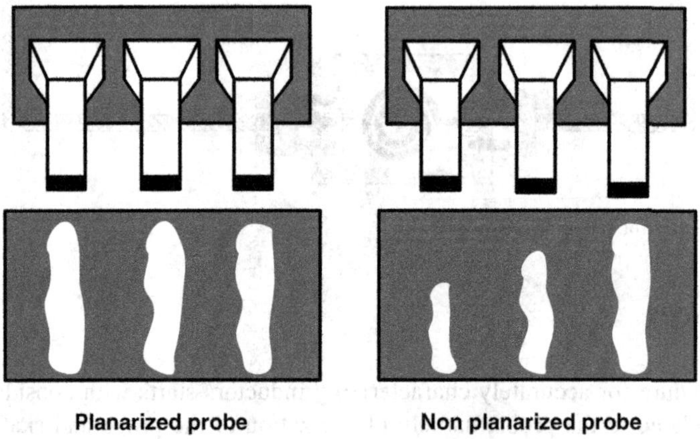

Fig. 7.3 Test on the contact substrate revealing the status of the probe in terms of planarization

Skating

When a coplanar probe touches down on the pad, the tip is normal to the wafer surface, but because the probe's body is at an angle to the wafer, lowering it more causes the tip to move across the pad, as Fig. 7.4 shows. This phenomenon is known as skating, and it is always necessary since pad metal layers may not be deposited with the same thickness on all wafers, and that may cause connectivity problems. Skating ensures good electric contact with the pad.

Skating is especially important when performing a VNA calibration or inductor measurements after calibration because the parasitic effects between the probe tip and the pad will depend on the applied skate. In order to perform accurate measurements it is important that the values of the probe tip parasitics remain constant. If the amount of applied skating is different from calibration to inductor measurements, the values for the probe tip parasitics will not remain constant, thus introducing uncertainty, especially at high frequencies. Therefore, it is essential

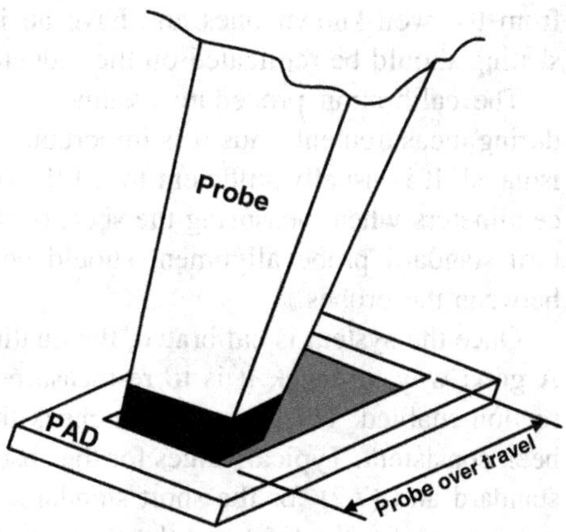

Fig. 7.4 Effect of skating on the pad

that the probe touches the pad with a consistent pressure from touch-down to touch-down. The pressure can be adjusted by applying a proper amount of over travel and skate by means of the probe station's z micrometer. To guarantee reliable contact, a minimum skate of 45 μm should be applied when using a tungsten probe over an aluminum pad [1].

Cleaning

After extended use, the probes accumulate small aluminum fragments, for example during the scrubbing action of the skating, which requires periodic cleaning. Probes can be cleaned with either compressed air or isopropyl alcohol applied with a swab [2]. When wiping, movement should be away from the probe tips; otherwise the tips may be damaged.

Calibration

The calibration procedure removes the systematic and drift errors that come from imperfections in the VNA, temperature changes, interference, etc. The calibration procedure consists of replacing the inductor to be measured with well-known calibration standards and making use of the results to solve the error parameters. In the case of on-wafer measurements, the well-known calibration standards are the ones contained in an impedance standard substrate (ISS). There are many calibration approaches, but the most extended one is the SOLT (Short-Open-Load-Thru) approach [1].

For correct calibration, it is very important that proper planarization of the probe has been performed beforehand. In addition, the skating should be adjusted by using the alignment marks; otherwise the parasitics of the standard may change from the well-known ones and have an impact on measurement accuracy. The skating should be replicated on the inductor measurements for accurate results.

The calibration procedure assumes that both probes are sufficiently isolated during measurement; thus it is important to guarantee that both ports are properly isolated. It is usually sufficient to lift the other probe in the air and move it a few centimeters when measuring the short or the open standards. In particular, for the thru standard probe alignment should be very accurate, including the distance between the probes.

Once the system is calibrated, the quality of the calibration should be checked. A good way to check it is to re-measure the calibration standards with the correction enabled. This way we can check that probe contact during calibration has been consistent. Typical values for the absolute S-parameters are (7.1) for the open standard and (7.2) for the short standard. Finally, (7.3) and (7.4) are the recommendations for the load and the thru standards, respectively [3].

$$|S_{11}| = 0dB \pm 0.05dB^{\circ} \tag{7.1}$$

$$|S_{11}| = 0dB \pm 0.1dB \tag{7.2}$$

$$|S_{11}| < -40dB \tag{7.3}$$

$$\begin{aligned} |S_{11}|, |S_{22}| &< -40dB \\ |S_{21}|, |S_{12}| &= 0dB \pm 0.02dB \end{aligned} \tag{7.4}$$

De-Embedding

Due to the difficulty of building probe tips that are small enough to directly measure the inductor, a dedicated on-wafer test fixture becomes necessary, as shown in Fig. 7.5. Several design rules can be followed in order to minimize the parasitic effects of the test fixture [1]. However, as the test fixture always influences the inductor measurements, some de-embedding procedure is required in order to extract the effect of the test fixture parasitics from the measurement.

The de-embedding process consists of removing the parasitics of the test fixture by first modeling the parasitics as S-parameters circuits, i.e. parasitic admittances and impedances. The whole set of the parasitics of the inductor test fixture can be seen in Fig. 7.6.

Once we have represented the test fixture by admittances and impedances, now several in-fixture standards are necessary in order to accurately characterize the parasitics introduced by the test fixture. The in-fixture standards needed to characterize the ten parameters of the test fixture are the single open, single short, open and short. Figure 7.7 shows the single open and single short in-fixture standard and Fig. 7.8 shows the specific parasitics that can be calculated from this in-fixture standard to further remove the parasitics from the measurements of the inductor

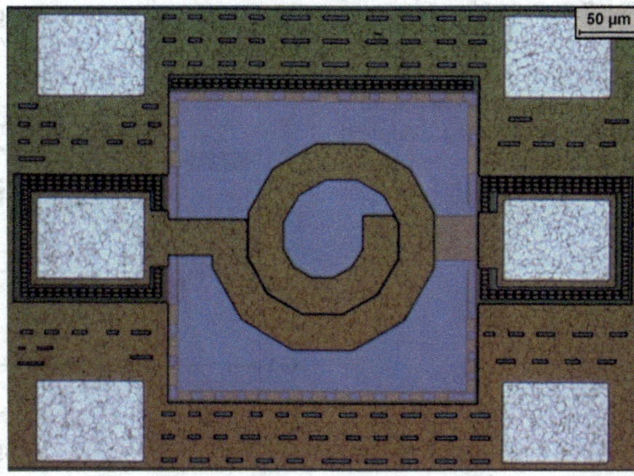

Fig. 7.5 Output stage drain inductor within its test fixture

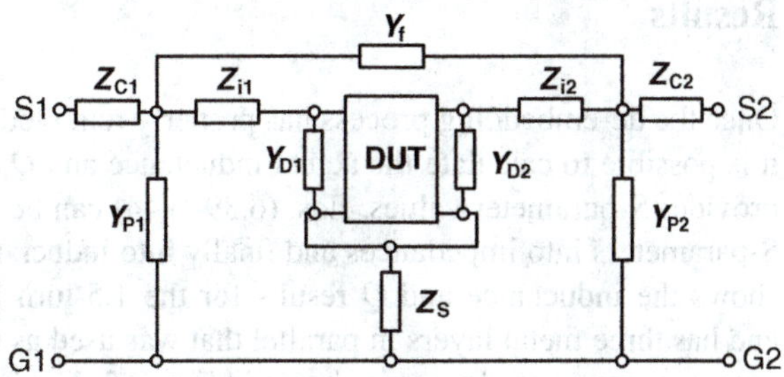

Fig. 7.6 Test-fixture model

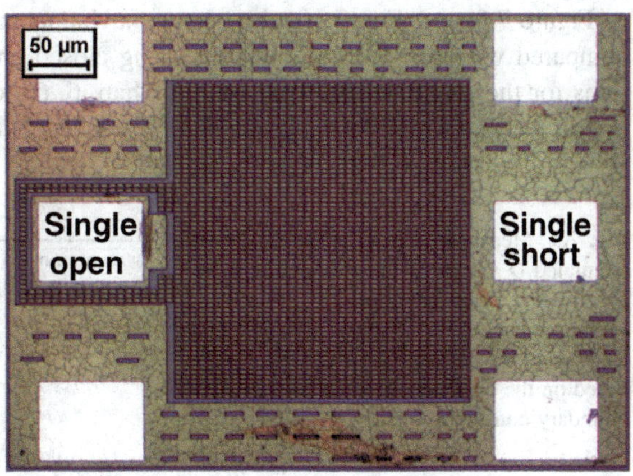

Fig. 7.7 Single open, single short in-fixture standard

within its test fixture. These specific in-fixture standards are based on the de-embedding procedure proposed in [3]. It must be noted that this de-embedding technique can be simplified if the test fixture is properly designed so that some of the parasitics can be considered negligible [1].

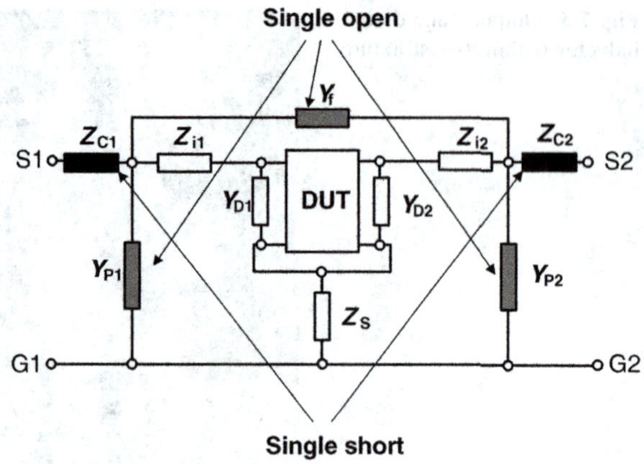

Fig. 7.8 Parasitics that are extracted by means of the single open, single short in-fixture standard (when both the open and short are applied to each probe)

Results

Once the de-embedding process has properly removed all the test fixture parasitics, it is possible to calculate the actual inductance and Q of the inductor. As the VNA provides S-parameter values, Eqs. (6.39–6.44) can be used in order to translate the S-parameters into impedances and finally into inductance and Q values. Figure 7.9 shows the inductance and Q results for the 1.5-turn inductor that is 28 μm wide and has three metal layers in parallel that was used as the drain inductor for the PA output stage, whose layout is shown in Fig. 7.5. The inductance results can be seen in Fig. 7.10.

Figure 7.9 and Fig. 7.10 also show the inductance and Q measurements as compared with the ADS simulations using closed and open boundary configurations for the substrate. As discussed in Chap. 6, the open boundary configuration that implements infinite substrate thickness achieves more accurate results, and in

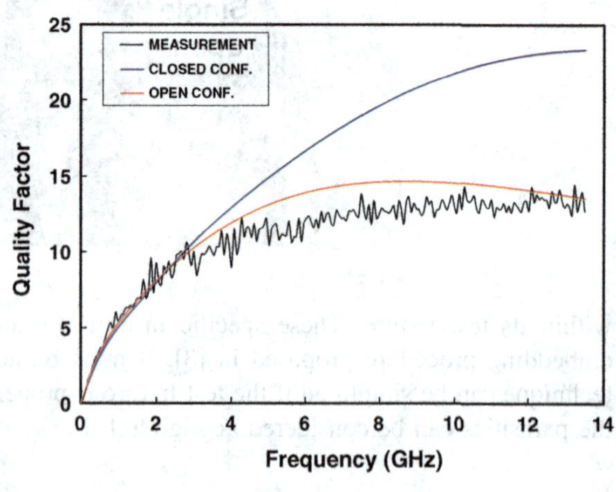

Fig. 7.9 Measured and simulated Q values of the output stage drain inductor. Measurements are compared to simulations performed based on the open and closed boundary configurations

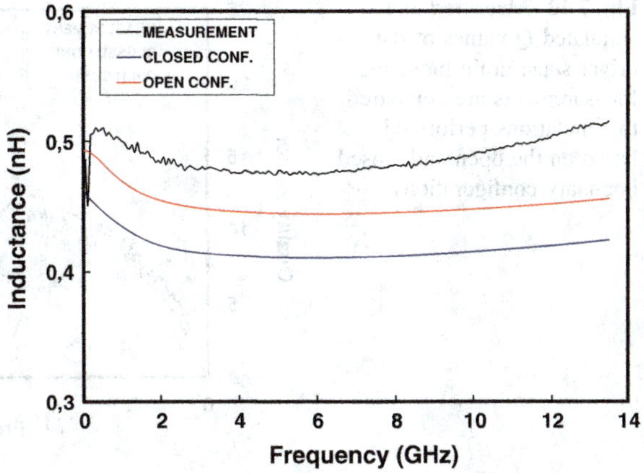

Fig. 7.10 Measured and simulated inductance values of the output stage drain inductor. Measurements are compared to simulations based on the open and closed boundary configurations

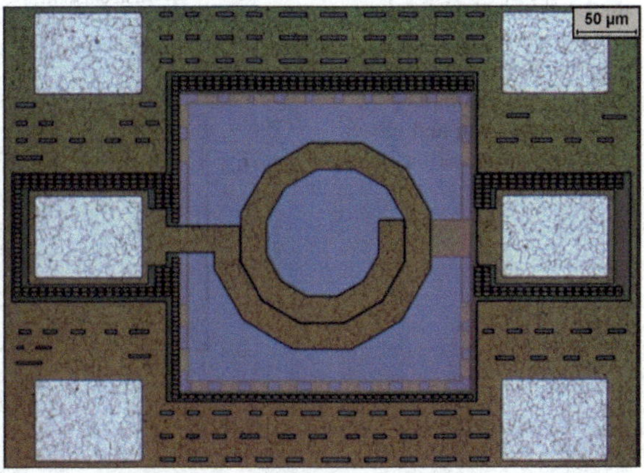

Fig. 7.11 Driver stage drain inductor within its test fixture

this case it's not only for the Q but also for the inductance results. This conclusion is also confirmed by the measurements of the drain inductor of the PA driver stage. This inductor, shown in Fig. 7.11, has been implemented with two metal layers in parallel, which are 20 lm wide, and have 1.5 turns. In addition, Figs. 7.12 and 7.13 illustrate the measured results compared to the ADS simulations.

The balanced inductor of the output matching network shown in Fig. 7.14 also provides improved results when the open boundary configuration is used, as can be seen in Figs. 7.15 and 7.16. This inductor is a balanced two metal layer inductor that has 2.5 turns and is 20 µm wide.

Finally, the balanced inductor of the input matching network has been also measured and compared with simulations. Figure 7.17 presents a 16-µm-wide inductor that is implemented using only the top metal layer. Its Q and inductance results are found in Figs. 7.18 and 7.19. In this case, the closed boundary configuration provides more accurate results. However, the simulation errors are low also for the open boundary configuration.

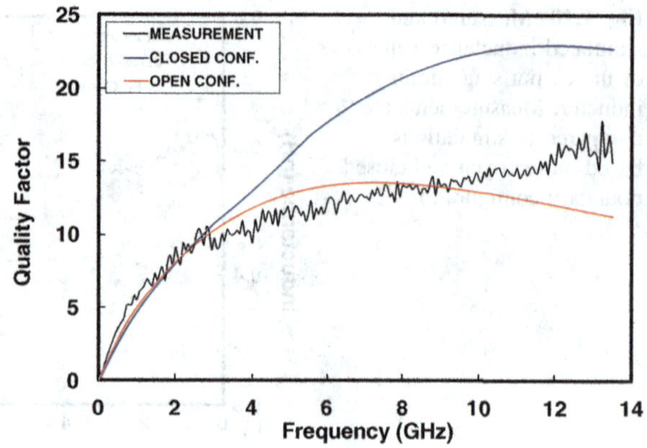

Fig. 7.12 Measured and simulated Q values of the driver stage drain inductor. Measurements are compared to simulations performed based on the open and closed boundary configurations

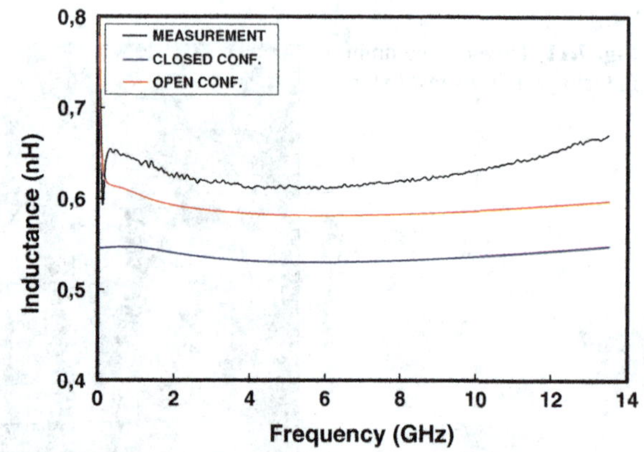

Fig. 7.13 Measured and simulated inductance values of the driver stage drain inductor. Measurements are compared to simulations based on the open and closed boundary configurations

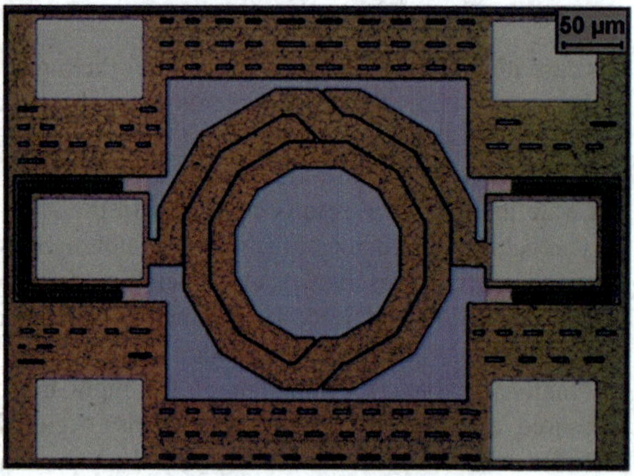

Fig. 7.14 Balanced inductor of the output matching network within its test fixture

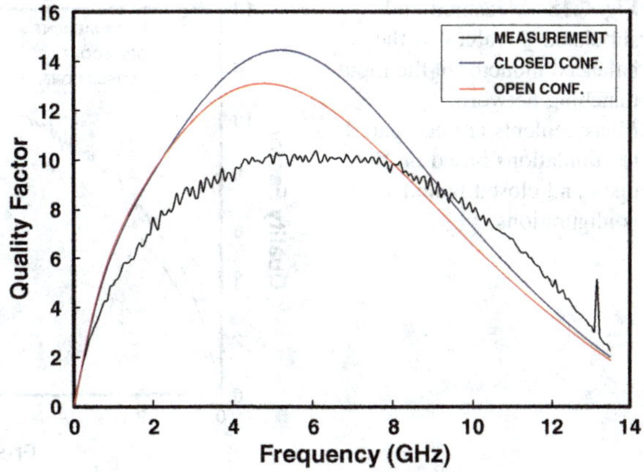

Fig. 7.15 Measured and simulated *Q* values of the balanced inductor of the output matching network. Measurements are compared to simulations based on the open and closed boundary configurations

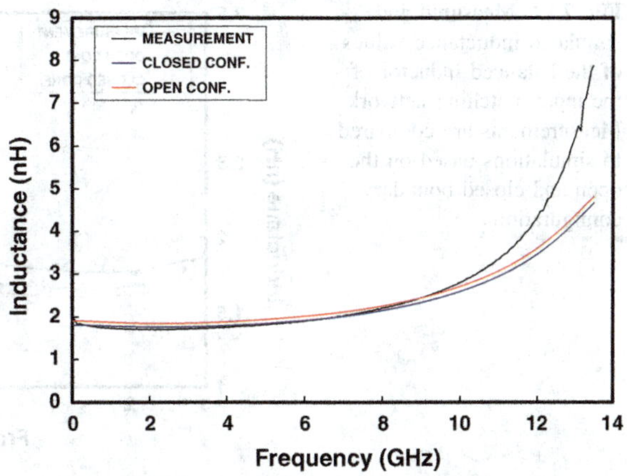

Fig. 7.16 Measured and simulated inductance values of the balanced inductor of the output matching network. Measurements are compared to simulations based on the open and closed boundary configurations

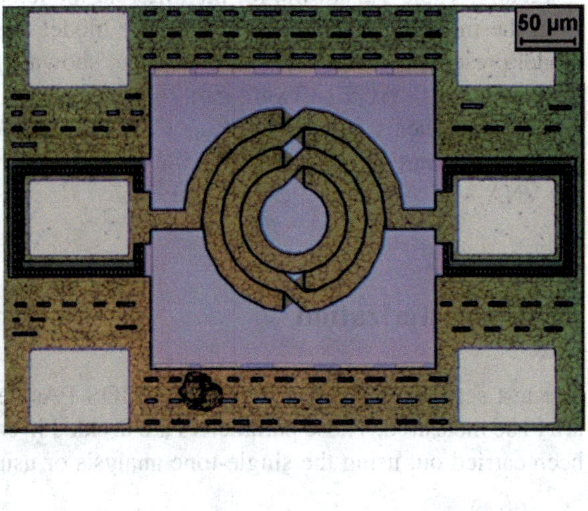

Fig. 7.17 Balanced inductor of the input matching network within its test fixture

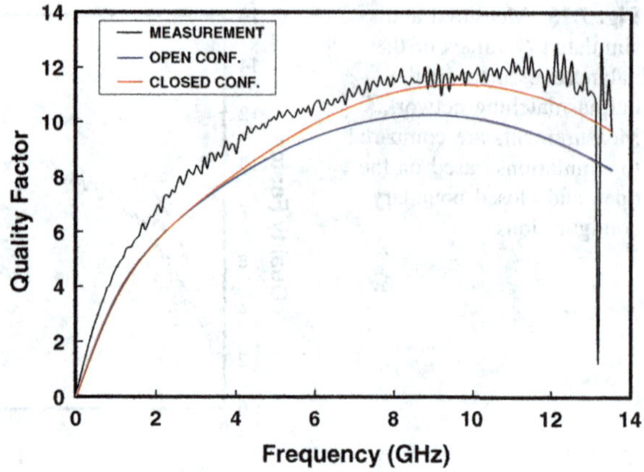

Fig. 7.18 Measured and simulated Q values of the balanced inductor of the input matching network. Measurements are compared to simulations based on the open and closed boundary configurations

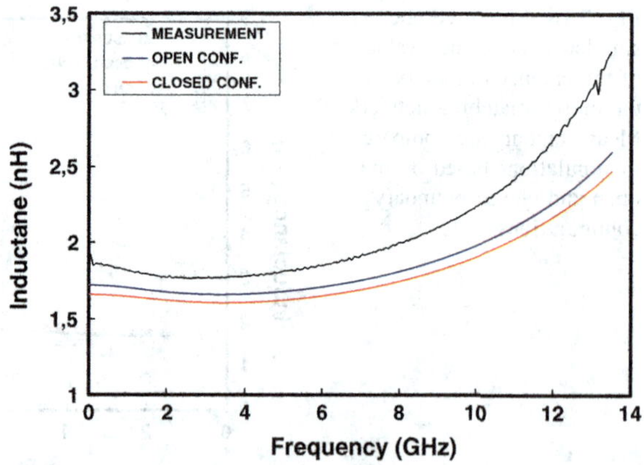

Fig. 7.19 Measured and simulated inductance values of the balanced inductor of the input matching network. Measurements are compared to simulations based on the open and closed boundary configurations

Finally, Table 7.2 shows, for the different inductors, the model values obtained from the measured S-parameters file. The model used for the inductor is the PI-model presented previously in Chap. 6 and shown again in Fig. 7.20. As we also saw in Chap. 6, ADS software can be used to obtain the different values of the model components (C_{par}, L_{ser}, R_{ser}, C_{ox}, C_{sub} and R_{sub}), although this time the values have been obtained from the file of the measured S-parameters provided by the VNA.

PA Characterization

The test setups for characterizing RF CMOS PAs depend on the parameters that are to be measured. These parameters are divided in two groups, whether they have been carried out using the single-tone analysis or using a digital channel analysis.

Table 7.2 Obtained values for the PI-model based on the measured results for inductors

Inductor lumped model

Parameter	Ind. D2 (Fig. 7.5)	Ind. D1 (Fig. 7.11)	Ind. OUT (Fig. 7.14)	Ind. IN (Fig. 7.17)
R_{ser} (Ω)	1.126	1.6	3.67	4.156
L_{ser} (pH)	471.9	594.7	1643	1715
R_{sub1} (Ω)	1038	1749	824.2	1926
R_{sub2} (Ω)	3266	5990	2393	2000
C_{ox1} (fF)	300	116.7	89.81	53.28
C_{ox2} (fF)	113.4	77.1	153.8	58.37
C_{sub1} (fF)	48.13	33.15	32.7	19.93
C_{sub2} (fF)	44.9	53.53	49.47	50.49
C_{par} (fF)	1.24	3.25	29.28	11.13

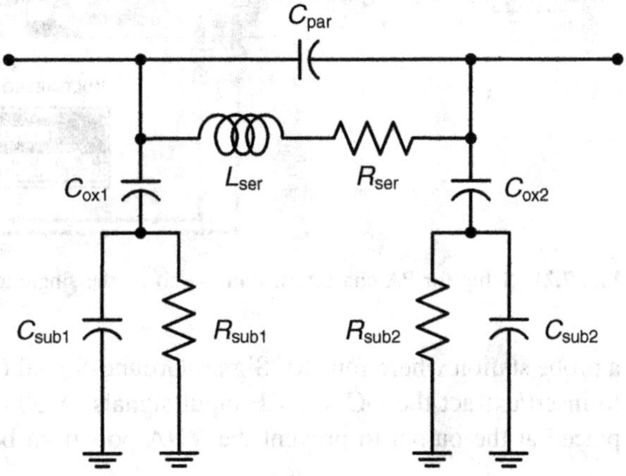

Fig. 7.20 The PI-model utilized for the inductors. C_{par} refers to the coupling capacitance between windings. L_{ser} is the inductance and R_{ser} the parasitic resistance. C_{ox} refers to parasitic capacitance to the substrate and C_{sub} and R_{sub} models the substrate

Single-Tone Tests

This section presents the procedure to follow in order to obtain the input matching, the power gain, and the current consumption, the P_{1dB}, the P_{SAT}, the PAE and the PA stability. All these parameters can be obtained by the so-called single-tone test.

Test Setup

A diagram of the test setup is shown in Fig. 7.21. The test requires a Vector Network Analyzer (VNA), the power supplies and digital multimeters. Additionally, the single/differential conversion for input and output signals is performed by means of two 180° hybrid couplers. Two DC blocks have been inserted at the input in order to decouple the gate bias of the CS transistors. The test for the IC requires

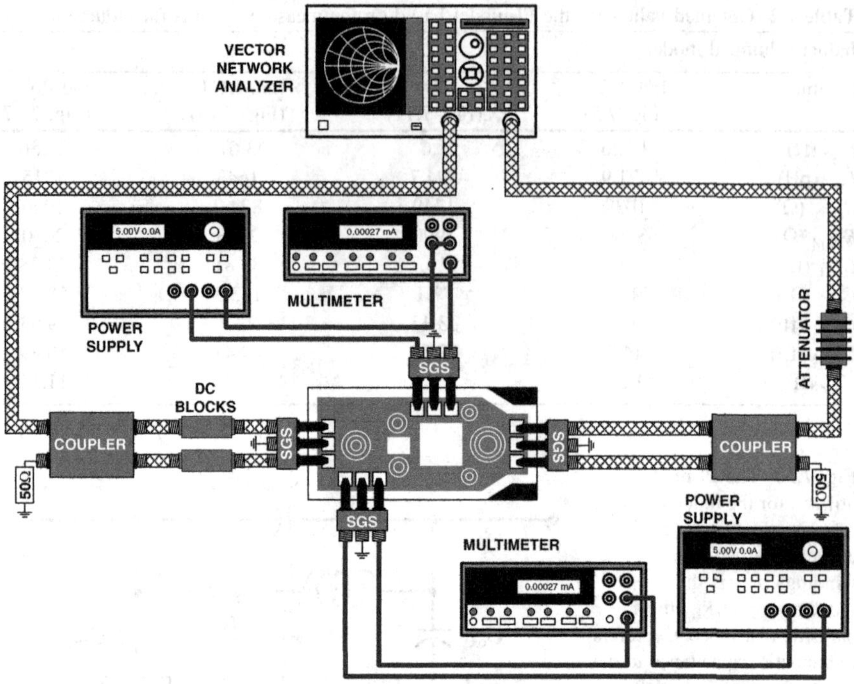

Fig. 7.21 Setup for PA characterization based on the single-tone test

a probe station where four RF Signal-Ground-Signal (SGS) probes have been used to insert/extract the DC and RF input signals. A 30 dB fixed attenuator has been placed at the output to prevent the VNA port from being damaged.

Input Matching and Gain

An S-parameter test is required to determine the input matching (S11) and gain (S21) of the PA. It must be ensured that the power level at which the test is performed is within the linear range of the PA. The fabricated PA presented in Chap. 6 has been measured following this setup. The results of the input matching and the gain are presented in Fig. 7.22.

Output P_{1dB} and P_{SAT}

The VNA can be also used for the P_{1dB} and P_{SAT} parameters, although in this case a power sweep is performed instead of a frequency sweep. The VNA is fixed to the frequency at which the measurement is to be taken, while the power is swept from the linear to the nonlinear regions of the PA. The test allows the P_{IN}-P_{OUT} and gain

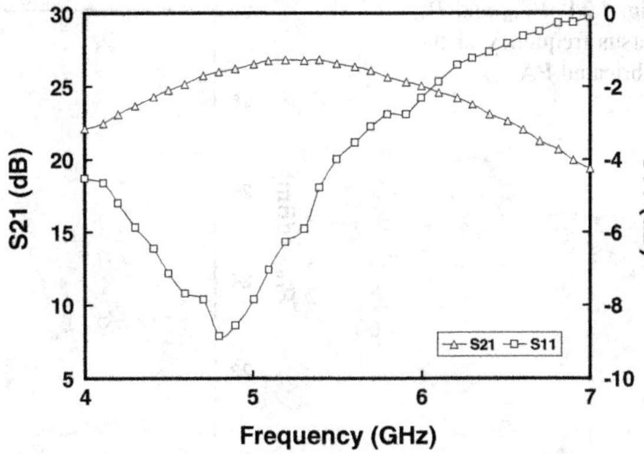

Fig. 7.22 Input matching (S11) and power gain (S21) results for the PA from an S-parameter test

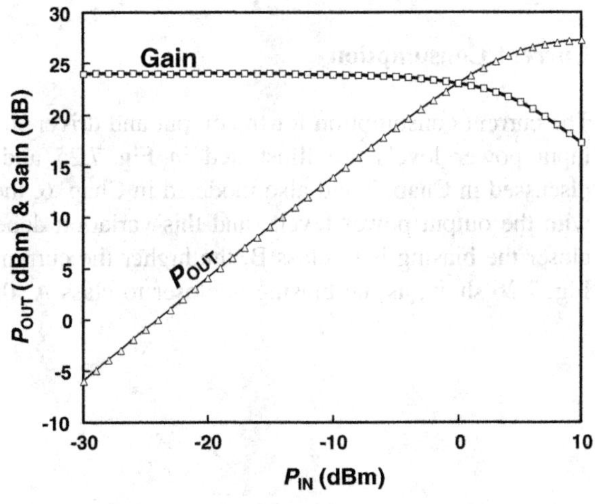

Fig. 7.23 P_{IN}-P_{OUT} and Gain results of the fabricated PA at 4.2 GHz

results of the PA to be obtained for a certain frequency. Figure 7.23 shows the P_{IN}-P_{OUT} and gain results of the fabricated PA at 4.2 GHz. In order to fully characterize the PA, this test must be performed for the frequency range at which the PA is intended.

From the P_{IN}-P_{OUT} or gain results it is possible to obtain the P_{1dB}, as it is the point at which the gain falls one decibel. Parameter P_{SAT} is obtained in the deep nonlinear region of the PA where only a residual power increase is observed at P_{OUT} for increases in P_{IN}, although the result for the P_{SAT} parameter should be accompanied by the amount of gain compression at which P_{SAT} has been measured. Since both the P_{1dB} and P_{SAT} parameters have been measured for different frequencies, it is possible to know the frequency range at which the PA works optimally, as Fig. 7.24 shows. As we discussed in Chap. 2, for a standard with stringent EVM requirement the parameter to focus on is the P_{1dB}.

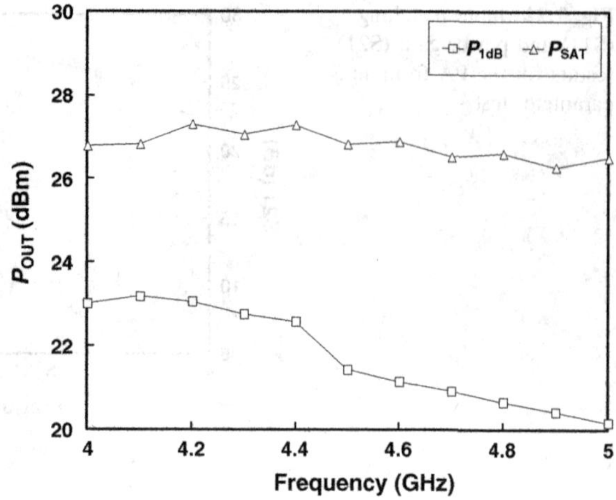

Fig. 7.24 P_{1dB} and P_{SAT} versus frequency of the fabricated PA

Current Consumption

The current consumption for the output and driver stages at 4.2 GHz and different input power levels are illustrated in Fig. 7.25 and Fig. 7.26, respectively. As discussed in Chap. 2 and also modeled in Chap. 6, the current consumption varies with the output power levels, and this variation depends on the PA biasing. The closer the biasing is to class B, the higher the current variation. As Fig. 7.25 and Fig. 7.26 show, as the biasing is closer to class A, the current is nearly constant.

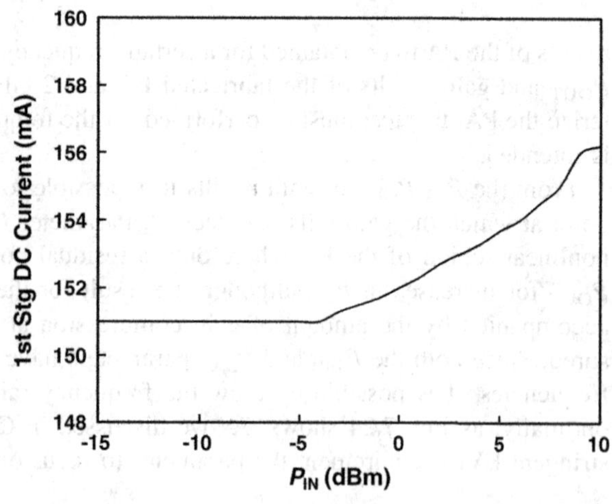

Fig. 7.25 Current consumption of the output stage of the fabricated PA at 4.2 GHz

Fig. 7.26 Current consumption of the driver stage of the fabricated PA at 4.2 GHz

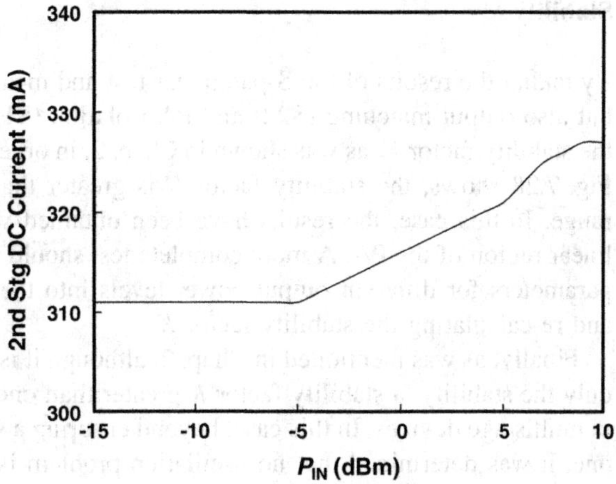

Power-Added Efficiency

The PAE can be obtained by taking the previous results of the P_{IN}-P_{OUT} and the current consumption values for each P_{IN} level and following (2.3). The results obtained at 4.2 GHz are shown in Fig. 7.27. It must be noted that for correct characterization, the P_{IN}-P_{OUT} and the current consumption results must be obtained for each frequency and the current consumption of the driver stage must be also considered.

Fig. 7.27 Power-added efficiency of the fabricated PA at 4.2 GHz

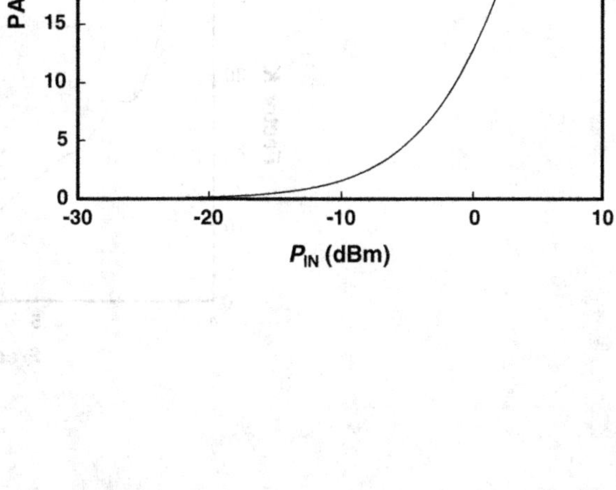

Stability

By taking the results of the S-parameter test and measuring not only S11 and S21 but also output matching (S22) and PA isolation (S12), it is possible to calculate the stability factor *K*, as was shown in Chap. 2, in order to quantify the stability. As Fig. 7.28 shows, the stability factor *K* is greater than one for the PA frequency range. In this case, the results have been obtained with power levels within the linear region of the PA. A more complete test should be done by measuring the S-parameters for different output power levels into the nonlinear region of the PA and re-calculating the stability factor *K*.

Finally, as was mentioned in Chap. 2, although it is a common practice to check only the stability, a stability factor *K* greater than one is not a sufficient condition in multistage devices. In this case, beyond ensuring a stability factor *K* greater than one, it was determined that no oscillation problem is observed at any frequency, which confirms the transient simulations performed during the design process.

Digital Channel Tests

This section presents the procedure to follow in order to obtain the maximum output power levels that comply with the spectrum emission mask and the EVM of a standard.

Test Setup

Tests based on a digital channel as an input can be performed by means of the setup in Fig. 7.29. This setup substitutes the VNA by a Signal Generator (SGN) and a Spectrum Analyzer (SPA). The SGN arbitrary wave generation capabilities

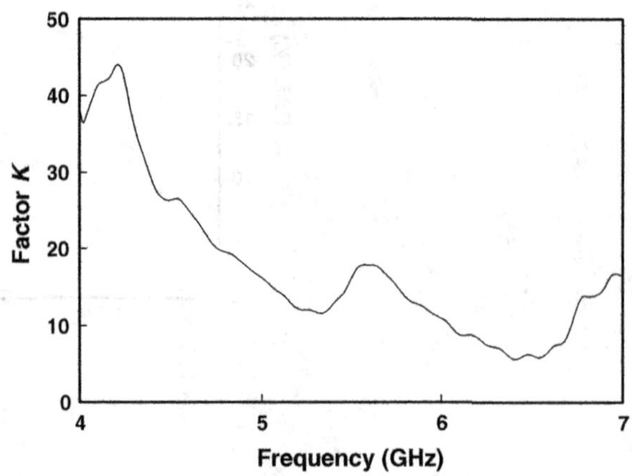

Fig. 7.28 Stability factor *K* of the fabricated PA

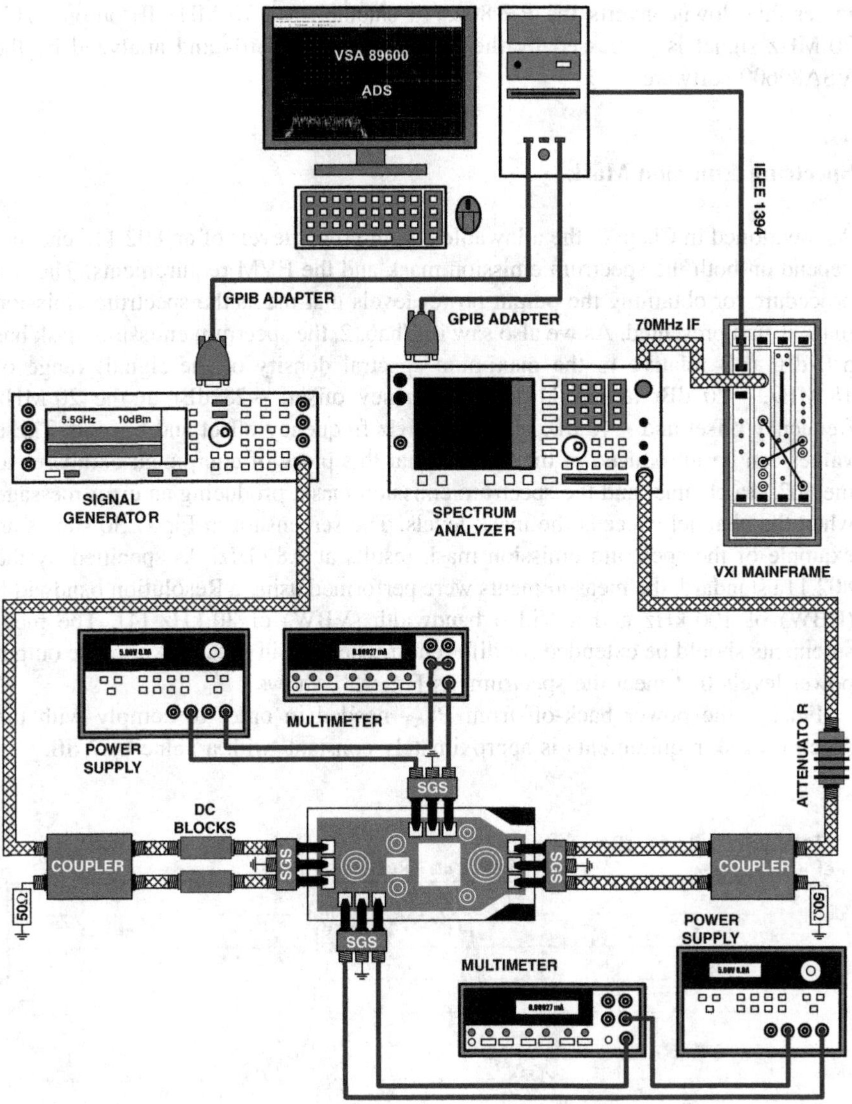

Fig. 7.29 Setup for PA characterization based on the digital channel test

are exploited by means of a connection to the ADS software. With ADS, a sample of a digital channel is created and downloaded to the SGN. In this case an 802.11a channel with 16 MHz bandwidth has been downloaded.

For the tests that determine the spectrum emission mask compliant measurements, the SPA capabilities are used in order implement the spectrum emission mask of the 802.11a standard and to measure the output power of the digital channel. In the case of EVM measurements, the Spectrum Analyzer is used as a mixer that downconverts the RF 802.11a channel to a 70 MHz IF output. This 70 MHz signal is processed by the VXI Mainframe cards and analyzed by the VSA89600 software.

Spectrum Emission Mask

As mentioned in Chap. 2, the allowable output power levels of an 802.11a channel depend on both the spectrum emission mask and the EVM requirements. The test procedure for obtaining the output power levels that meets the spectrum emission mask is first presented. As we also saw in Chap. 2, the spectrum emission mask has a 0 dBr (dB relative to the maximum spectral density of the signal) range of 18 MHz, −20 dBr at the 11 MHz frequency offset, −28 dBr at the 20 MHz frequency offset and −40 dBr at the 30 MHz frequency offset and beyond. These values can be introduced in the SPA so that this piece of equipment compares to the 802.11a channel and the spectrum emission mask, producing an error message when the channel exceeds the mask levels. The screenshot in Fig. 7.30 shows an example of the spectrum emission mask results at 4.8 GHz. As specified by the 802.11a standard, the measurements were performed using a Resolution bandwidth (RBW) of 100 kHz and a Video bandwidth (VBW) of 30 kHz [4]. The measurements should be extended for different frequencies in order to know the output power levels that meet the spectrum, as Fig. 7.31 shows.

Finally, the power back-off from P_{SAT} needed in order to comply with the spectral mask requirements is approximately constant, with a value of 7 dB.

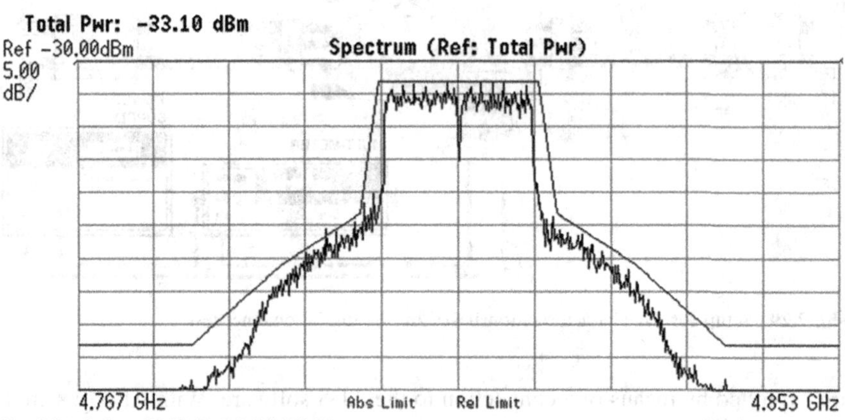

Total Pwr Ref: -33.10 dBm / 400.000 kHz

			Lower		<-Peak->	Upper
Start(Hz)	Stop(Hz)	Meas BW(Hz)	dBm	Freq(Hz)	dBm	Freq(Hz)
0.0000	9.0000 M	100.00 k	-33.26	4.7988 G	-33.16	4.8038 G
9.0000 M	11.000 M	100.00 k	-50.11	4.7904 G	-51.58	4.8091 G
11.000 M	20.000 M	100.00 k	-54.50 F	4.7885 G	-55.82	4.8112 G
20.000 M	30.000 M	100.00 k	-63.49	4.7800 G	-63.54	4.8200 G
30.000 M	43.000 M	100.00 k	-78.51	4.7699 G	-78.35	4.8300 G

Fig. 7.30 Screenshot of the spectrum emission mask results at 4.8 GHz

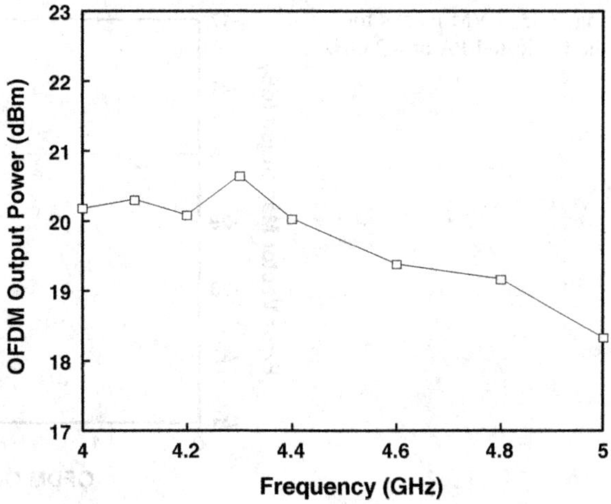

Fig. 7.31 Output power levels complying with the spectrum emission mask requirements of the 802.11a standard at different frequencies

Error Vector Magnitude

The measurements of the EVM require demodulating the channel in order to measure modulation accuracy. This task is performed with the VSA 89600 software package, using the IF output signal provided by the SPA. A screenshot of the software is in Fig. 7.32.

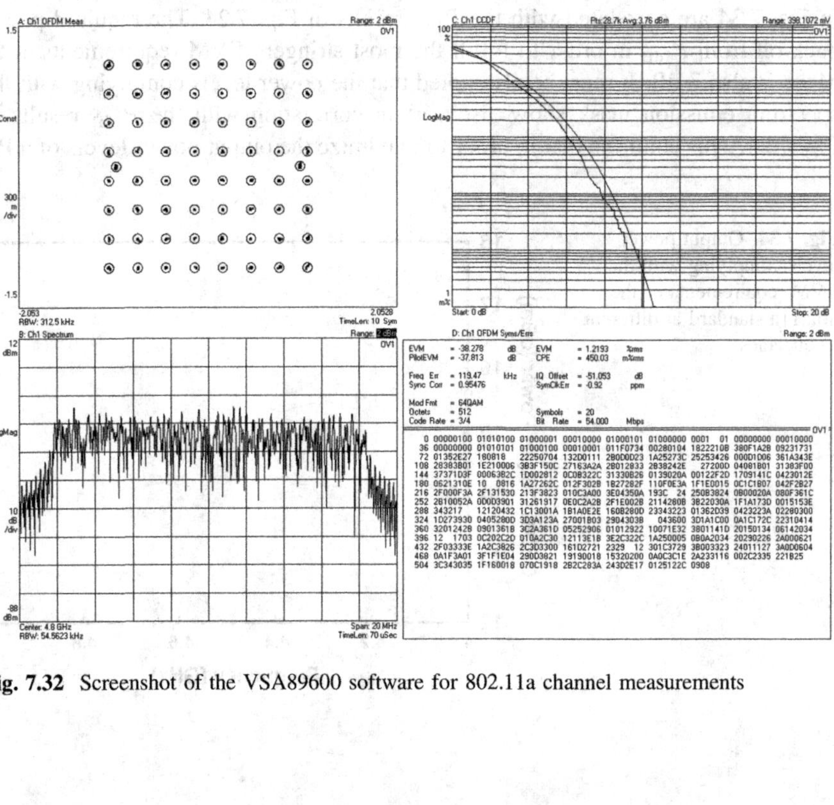

Fig. 7.32 Screenshot of the VSA89600 software for 802.11a channel measurements

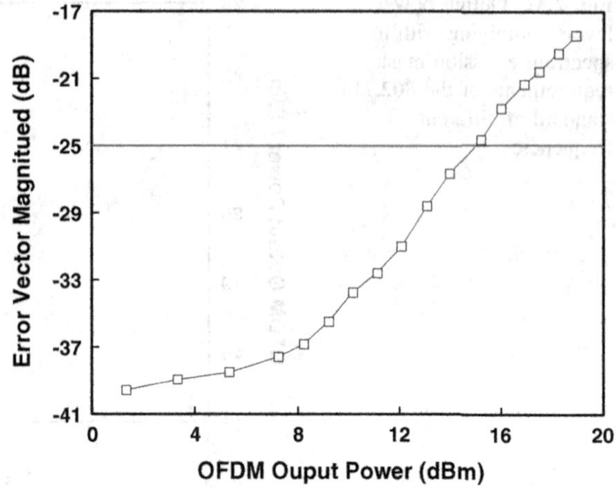

Fig. 7.33 EVM results for the fabricated PA at 4.2 GHz

The results in Fig. 7.33 show the complete EVM test for the PA at a frequency of 4.2 GHz. The input power of the 802.11a channel is increased and the value of the EVM is taken for each input power value. As the PA enters the nonlinear region, the EVM is degraded. The solid red horizontal line in Fig. 7.33 indicates the EVM value in dB (−25 dB), which must not be exceeded at the highest bit rate of 54 Mbps.

In addition, Fig. 7.34 shows the output power levels of the PA while complying with the EVM at 54 Mbps at different frequencies. As can be observed, the results in Fig. 7.34 are correlated with the P_{1dB} results of Fig. 7.24. The required power back-off from P_{1dB} in order to fulfill the most stringent EVM requirements at 54 Mbps is also 7 dB. It must be also noted that the power levels complying with the spectrum emission mask show also certain correlation with the P_{1dB} results in Fig. 7.24. This confirms that in order to maximize the output power levels of a PA

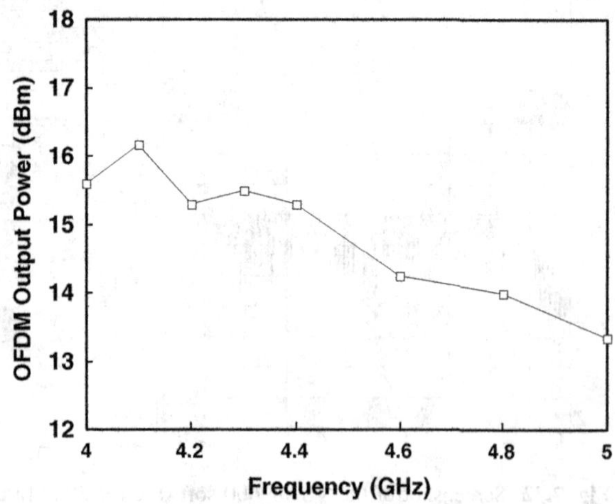

Fig. 7.34 Output power levels complying with the EVM requirements of the 802.11a standard at different frequencies

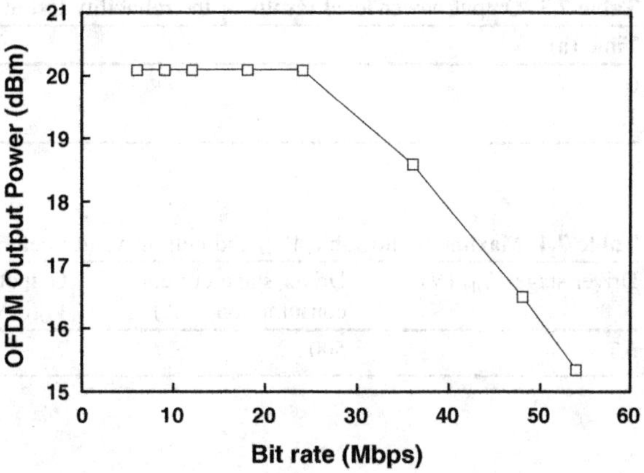

Fig. 7.35 Output power levels complying with the EVM and the spectrum emission mask requirements at 4.2 GHz and all the 802.11a standard bit rates

intended for a standard with stringent EVM requirements, P_{1dB} is the parameter that must be optimized.

A further study is to know at which bit rate the EVM requirements start to be dominant in the channel output power levels. As mentioned in Chap. 2, the EVM requirements at 54 Mbps are the limiting ones. However, given that the EVM requirements relax as the bit rate decreases, not all the bit rates will be EVM limited. In fact, the measurement results presented in Fig. 7.35 shows that the EVM requirements stop being dominant at 24 Mbps. At this bit rate and the lower ones, it is only the spectrum emission mask that limits the output power level.

Reliability and Maximum Rating

The reliability study should be carried out at the maximum power levels the PA will work with. In this case, the reliability test has been performed with the PA working at an output power level that is only limited by the spectrum emission mask. In this test, the PA must maintain its performance while working uninterruptedly at the greatest output power level. The results presented in Table 7.3 were taken at the beginning of the test and after 15 h of working. As mentioned in Chap. 4, any PA degradation becomes evident during the first few hours of continuous operation [5, 6]. Table 7.3 shows that the output power level stays almost constant after 15 h of working and therefore no degradation is appreciated.

In addition, the maximum rating test was performed by increasing the supply voltage of both the driver and the output stage and measuring the current consumptions of these stages. Table 7.4 shows the maximum supply voltages and current consumptions that were reached before an overall PA failure due to the excessive current flow through the metal tracks, which causes them to melt. This test shows that normal conditions are far from the limits at which the PA fails.

Table 7.3 Output power level results of the reliability test at 4.2 GHz

Time (h)	Output power (dBm)
0	19.82
15	19.74

Table 7.4 Maximum allowable V_{DD} and current values before PA failure

Driver stage V_{DD} (V)	Driver stage current consumption (mA)	Output stage V_{DD} (V)	Output stage current consumption (mA)
6.7	500	7	1000

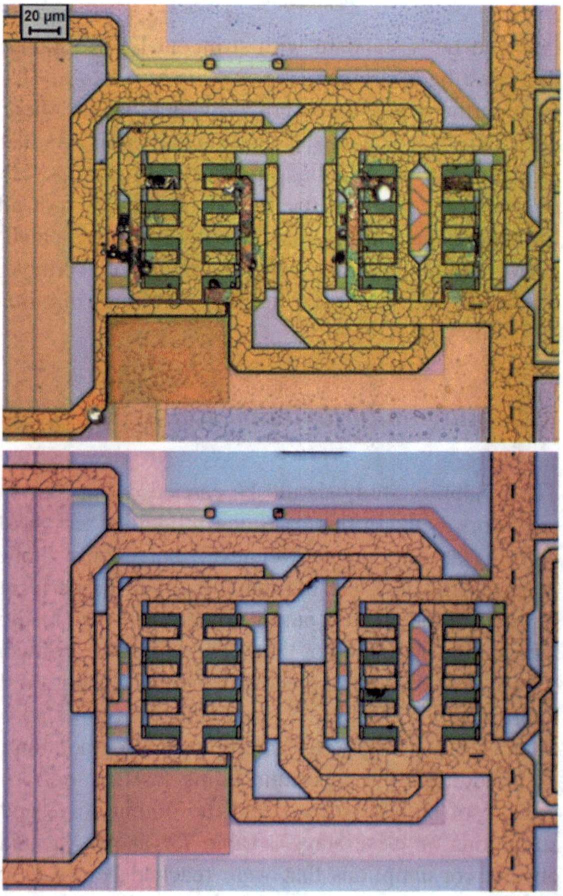

Fig. 7.36 Microphotographs of the driver stage core showing the effects of excessive current in the drain transistor metal tracks (*top*) and unaffected driver stage core (*bottom*)

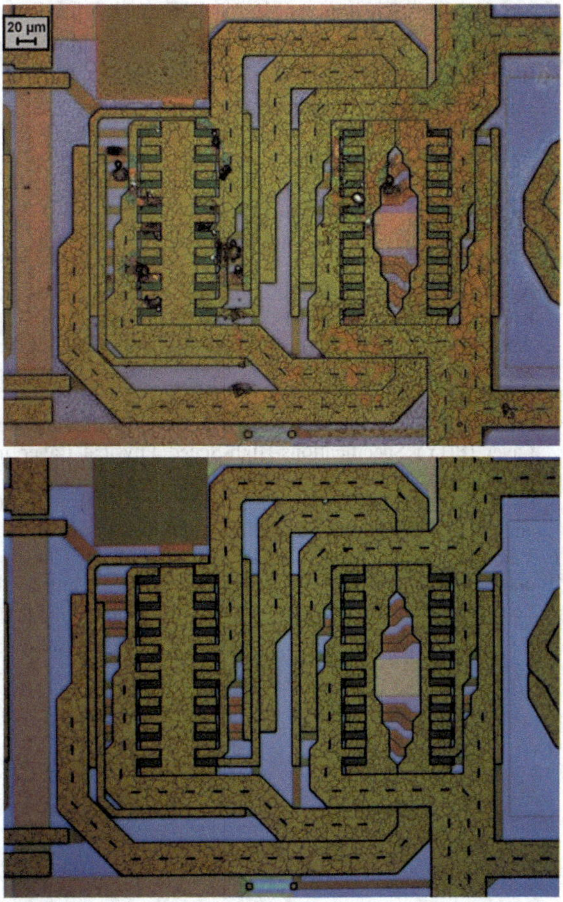

Fig. 7.37 Microphotographs of the output stage core showing the effects of excessive current in the drain transistor metal tracks (*top*) and unaffected output stage core (*bottom*)

Finally, Figs. 7.36 and 7.37 show photographs of the driver and output stage cores before and after maximum rating measurement. Note that the metal tracks appear melted mainly at the drains of the common-source transistors.

Conclusions

This chapter presented the equipment, test setups, procedures and practical recommendations for accurately characterizing integrated inductors intended for PAs. Specifically, the Q and inductance results were presented for several inductors and then compared to simulations so that the best configuration for the substrate used in simulations can be chosen. We also detailed the test setups and procedures for a complete characterization of a fully integrated linear CMOS PA. The test can be separated into two different groups: single-tone tests and digital channel tests.

From those tests it is possible to characterize the linearity and efficiency of the PA based on the most common parameters, namely P_{1dB}, P_{SAT}, PAE, spectrum emission mask and EVM. Finally, results related to stability, reliability and maximum rating tests were also presented.

References

1. Aguilera J, Berenguer R (2003) Design and test of integrated inductors for RF applications. Kluwer Academic Publishers, The Netherlands
2. Wartenberg S (2003) RF coplanar probe basics. Microwave J 46(3):20–38
3. Kolding TE (2000) RF on-wafer device measuring techniques. Aalborg University Aalborg, Denmark
4. IEEE Std 802.11a-1999. Part 11: Wireless LAN Medium Access Control (MAC) and Physical Layer (PHY) Specifications-High-Speed Physical Layer in the 5 GHz Band, IEEE, 2000. http://standards.ieee.org. Accessed 1 Apr 2013
5. Sowlati T, Leenaerts DMW (2003) A 2.4– GHz 0.18 mm CMOS self-biased cascode power amplifier. IEEE J Solid-State Circuits 38(8):1318–1324
6. Vathulya VR, Sowlati T, Leenaerts D (2001) Class 1 bluetooth power amplifier with 24 dBm output power and 48 % PAE at 2.4 GHz in 0.25 mm CMOS. In: Proceedings of European Solid-State Circuits Conference (ESSCIRC 2001) Austria, pp 57–60

Chapter 8
Conclusion

Abstract This chapter concludes the book by highlighting the main aspects that have been discussed in the field of RF CMOS linear power amplifiers. As mentioned throughout the book, there is great interest shown in CMOS integration of linear PAs for wireless communication systems. However, as CMOS processes present important issues to high-performance PAs, it is crucial to know how to overcome these issues by a proper design and implementation of PAs. A high performance RF CMOS PA is of great importance not only because a complete integration of the transceiver is then possible but also because of the great impact of the PA power consumption.

Highlights

This book treated all the important aspects in the design of CMOS linear RF power amplifiers.

Chapter 1 introduced the PA as a crucial block of a transceiver in the context of modern communication systems. The impact of the PA was quantified for several wireless standards showing different communication ranges. The main mobile communication standards were also considered in order to understand how the latest standards impose more stringent requirements due to the need of higher spectral efficiencies.

Chapter 2 presented the most usual metrics for PA performance characterization. These metrics can be arranged in two main groups; linearity metrics and efficiency metrics and they can differ if the input signal is just a single sinusoidal tone or a digital channel of a specific standard. In this chapter the concept of power back-off for a specific communication standard was also discussed and related to the common $P_{1\text{dB}}$ and P_{SAT} parameters of a PA.

Chapter 3 described the different PA classes; from the current source-type class A to class C PAs to the switch-type class D and class E PAs. Class F PAs were treated separately as PA class between current source and switch-type amplifiers.

H. Solar Ruiz and R. Berenguer Pérez, *Linear CMOS RF Power Amplifiers*,
DOI: 10.1007/978-1-4614-8657-2_8,

In addition, practical implementations of these PA classes within the state of the art were also discussed.

Chapter 4 dealt with the specific issues of CMOS process for PA implementation. The reliability issues of RF CMOS PA were detailed in this chapter. A low breakdown voltage avoids high supply voltages and so the CMOS processes impose a crucial limit for high linearity, high efficiency PAs along with high output power levels. In addition, the low supply voltage problem is aggravated in the scaling-down of the CMOS technology because relatively high knee voltage further limits the output voltage headroom.

Chapter 5 covers the most important techniques for enhancing the performance of linear PAs. Whenever a linear CMOS PA performance is not enough for a specific communication system, either because the linearity or the efficiency must be improved or because greater output power levels are required, these techniques allow a designer to improve one or several of these three characteristics. This chapter detailed simple techniques, as the use of the cascode transistor, analog predistortion, simple parallel combination or dynamic biasing as well as more complex architectures such as the transformer power combining, the Doherty architecture, digital predistortion or envelope tracking. For all these enhancement techniques their advantages, drawbacks and examples of practical implementations were presented and discussed.

Chapter 6 treated the design flow that a designer should follow in order to design a fully integrated linear CMOS PA: from the first PA specifications to the layout design. This chapter also described a model that proved very useful as a starting point for the PA design. The model takes into account effects such as PA biasing, current consumption, supply and knee voltages, inductor quality factors or the power gain in the P_{IN}-P_{OUT} and PAE-P_{IN} curves. In addition, this chapter also discussed the requirements that the integrated inductor must fulfill in order to be implemented in the output stages of a PA. Due to their specific requirements they were referred to as power inductors. The design flow that must be followed and the best geometry of these inductors were also explained in this chapter.

Chapter 7 dealt with the setups for the characterization of a fabricated PA and the implemented inductors. The setups were grouped depending on the type of parameters to be extracted. Single-tone test for parameters like the P_{1dB}, P_{SAT}, power gain or PAE and digital channel tests for the spectrum emission mask and the EVM. Characterization of the PA in terms of reliability was also presented along with the setups for characterizing the integrated inductors. A fabricated RF CMOS PA and its integrated inductors were taken as an example in order to illustrate the measured results following the described setups.

Main Contributions

This section presents the main contributions of the book.

A Complete Design Flow for a CMOS Linear PA

The work done in this book establishes the complete design flow for the optimization of linear CMOS power amplifier starting with the specifications that modern communication standards impose on PAs, starting with the first steps of the PA design to the final IC schematic and layout and the required test setups.

A model for linear power amplifiers facilitates the first design steps in terms of transistor sizing, required inductor quality factors or minimum supply voltage. The model considers the limitations that CMOS processes impose on the implementation of power amplifiers, as for example the knee voltage effect. As a result, the model provides the expected P_{IN}-P_{OUT} and PAE-P_{IN} curves fitting well with measured results of fabricated PAs.

Useful Enhancement Techniques for Linear CMOS PAs

The book proposes specific techniques and architectures that allow a designer to enhance the performance of CMOS linear PAs in terms of efficiency, linearity and output power levels. In addition, the advantages and drawbacks along with practical implementations from the state of art are presented and discussed.

Design and Characterization of Power Inductors

The integrated inductors implemented in a PA require special attention because several extra design considerations are necessary due to the special characteristics of the application. This fact has led to the new concept of power inductors. The design flow and practical advices that these inductors must follow in order to fulfill the extra design considerations are covered in this book. Finally, the required setups for the proper characterization of these inductors have been also covered.

Index

H. Solar Ruiz and R. Berenguer Pérez, *Linear CMOS RF Power Amplifiers*,
DOI: 10.1007/978-1-4614-8657-2,

9781461486565